THE ANCIENT EARTHWO
CRANBORNE CHASE

ALAN SUTTON &
WILTSHIRE COUNTY LIBRARY
1988

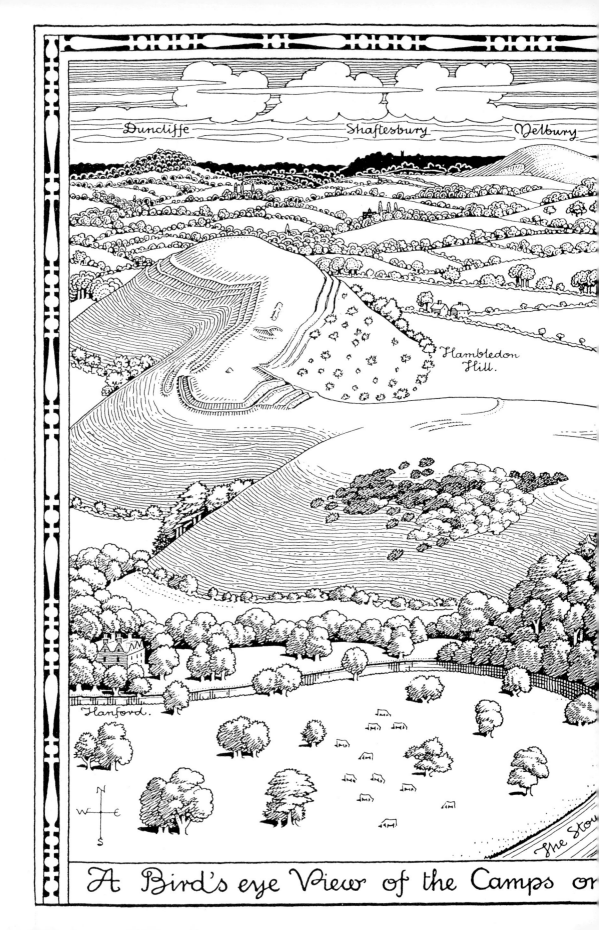

Duncliffe · Shaftesbury · Melbury

Hambledon Hill.

Hanford.

The Stou

A Bird's eye View of the Camps on

THE ANCIENT EARTHWORKS OF CRANBORNE CHASE

Described, & Delineated in plans founded
on the 25 inch to 1 mile Ordnance Survey
by

Heywood Sumner. F.S.A.

With a Map shewing the physical features
& the ancient sites of Cranborne Chase ♭
founded on the 1 inch to 1 mile Ordnance
Survey & coloured by hand ♭ A.D. 1913

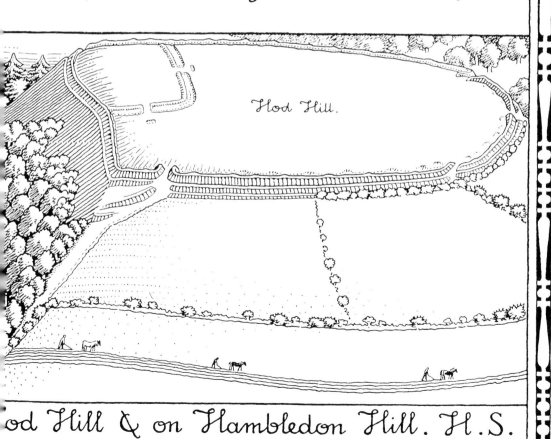

Hod Hill.

od Hill & on Hambledon Hill. H.S.

Published jointly by Alan Sutton Publishing Limited and Wiltshire County Library.

Alan Sutton Publishing Limited
30 Brunswick Road
Gloucester GL1 1JJ

First published 1913
This edition published 1988

British Library Cataloguing in Publication Data

Sumner, Heywood, *1853–1940*
    The ancient earthworks of Cranborne Chase.
    1. Dorset. Cranborne Chase. Earthwork
    structures : Curses
    I. Title
    936.2′334

    ISBN 0-86299-536-1

Cover Illustration A Bird's Eye View of the Camps on Hod Hill and on Hambledon Hill
by Heywood Sumner

Origination by
Alan Sutton Publishing Limited.
Printed in Great Britain.

HEYWOOD SUMNER
AND
THE ANCIENT EARTHWORKS OF
CRANBORNE CHASE

I N January 1904 Heywood Sumner and his family moved into their newly-built house at Cuckoo Hill, near Gorley on the western edge of the New Forest. For Sumner it was a momentous occasion: he was now 51 and the move symbolized the end of an era. He was giving up his life as a successful London artist and immersing himself in the quiet of the countryside. In London he had achieved a considerable success as a highly versatile member of the Arts and Crafts movement, his skills leading him to practise in a variety of media: book illustrations, sgraffito work, stained glass, wall paper designs, posters, tapestry and furniture. Now he was deliberately turning his back on success and looking forward to his new life as a solitary countryman.

His house at Cuckoo Hill was a labour of love. The Sumners had left London in 1897 for Bournemouth where he completed a series of artistic commissions while searching for an ideal place to settle. On the edge of the Forest he found it: 'a squatter's holding on the gravel hills of the Forest, where we could try the experiment . . . of life amid wild air'. The formalities completed, Sumner set about designing and overseeing the building of the family house. The foundations were laid in May 1902 and the family moved in on a frosty January night eighteen months later helped by the local villagers who turned out in force when they saw the removal carts bogged down in the muddy lane.

Sumner's early life had been uneventful. He was born in Old Alresford in 1853, son of the local Rector. His family were intimately involved in church affairs and his mother, Mary Elizabeth, was founder of the Mothers Union. After reading Law at Oxford, and qualifying as a barrister at Lincoln's Inn, he abandoned his profession to become a book illustrator, soon to broaden his interests to embrace the full range of applied art as befitted a follower of William Morris. For a brief period of barely twenty years he was to immerse himself in his artistic career, becoming master of many media. Why he left

CRANBORNE CHASE

London and retired to the countryside, when so comparatively young, we are unlikely ever to know. He tells us that it was because of his wife's failing health but the real reasons were more deep rooted. He was a solitary person searching for deeper truths: the brashness and superficiality of London society had no appeal for him. He had enjoyed his brief visits to the New Forest. Its wildness and calm offered a sympathetic setting for his innate pantheism to express itself and it was here that he determined to spend the rest of his life.

The transition period from retired artist to field archaeologist took about five years during which time he was busy observing and recording every aspect of his new environment in a beautifully illustrated scrap-book of which three volumes survive. The first of these, covering the period from 1904–10, he published under the title *The Book of Gorley* – a delightful miscellany of country life seen through the whimsical eye of the lone artist. By the time the book appeared he was fast developing a fascination for archaeology. From Cuckoo Hill he could look across the Avon valley to the great chalk upland of Cranborne Chase rich in magnificently-preserved prehistoric earthworks. Several of the major sites had been excavated by General Pitt-Rivers in the 1880s adding a new depth of meaning to these mysterious landscape features.

The years immediately before the First World War saw the birth of field archaeology in Britain. In 1906 the Committee on Ancient Earthworks and Fortified Enclosures of the Congress of Archaeological Societies had published a report calling for the systematic survey of all earthworks in the country. Many people had responded including the Hampshire doctor J.P. Williams-Freeman who, after eight years of painstaking field-work, published a complete survey of Hampshire earthworks in a remarkable book entitled *Field Archaeology as illustrated by Hampshire* (1915). Williams-Freeman and Sumner were good friends and often went on expeditions together and it was in this way that Sumner learnt the art of field surveying.

By the spring of 1911 he was ready to set out by himself to record the upstanding monuments of the Chase. Since the majority of his plans are dated his rapid progress across the countryside can be charted. By the winter of 1912–13 the field-work was complete and *The Ancient Earthworks of Cranborne Chase* appeared later in 1913. The book is a masterpiece: its meticulous plans are minor works of art, while the descriptions are models of clear observation and precise recording.

Today the book has taken on a new importance. Sumner's field-work took place at a time when much of the Wessex chalkland was still sheep pasture and the earthworks stood sharp in their landscape settings. Now most of the area has been reduced to a uniform prairie in the interests of intensive crop production. Though the more important of the individual sites remain unploughed, their settings and many minor earthworks have been obliterated. Sumner's survey was just in time. It records a landscape of immense antiquity on the threshold of destruction. For this reason Sumner's Cranborne Chase is, and will continue to be, a primary archaeological source. But it is much more than that – it is a book of great charm bristling with joy for the countryside: you can almost smell the aromatic chalkland turf and feel its spring beneath your feet. It shows its author – countryman, archaeologist and artist – at his best, content in a landscape he loved.

Cranborne Chase was Sumner's first serious sortie into archaeology. It was soon followed by a companion volume *The Ancient Earthworks of the New Forest* (1917) and by a spate of papers dealing with individual sites. Nor was Sumner satisfied simply to measure monuments: in 1911, at the age of 58, he began his first excavation on the Romano-British enclosure on Rockbourne Down and for the next 14 years he was to excavate and publish 13 sites almost single handed. In 1925, at the age of 72, he decided to retire from digging but until his death in 1940 he maintained a lively interest in field-work, wandering for long hours across heaths and through his beloved Forest in search of new earthworks and writing delightful accounts for local journals.

As an artist Heywood Sumner spanned the transition from Pre-Raphaelite to Art Nouveau, as an archaeologist he bridged the divide from the antiquarian to the professional field-worker. He justly deserves the modest recognition which the reprinting of this book will bring.

Barry Cunliffe
Oxford
June 1988

PREFACE

THIS book of plans is an attempt to put into practice the precepts of the Committee on Ancient Earthworks appointed by the Congress of Archaeological Societies—namely, that Plans and Schedules of our Ancient Earthworks should be made throughout England; that a definite area should be undertaken by each worker; and that such plans should be founded on the 25 inch Ordnance Survey. My daily view extends over Cranborne Chase, and curiosity had often led me to investigate its varied earthworks. In so doing I had felt the want of a complete record of their plans; thus it came to pass that two years ago I ventured to undertake a definite survey of the Ancient Earthworks on Cranborne Chase, the results of which are here set forth.

Most of these Earthworks may be found delineated in the excellent 6 inch and 25 inch Ordnance Survey Maps, but a few have been overlooked by the surveyors, while in several cases details of archaeological importance have been omitted. These oversights and omissions I have tried to supplement.

The area of Cranborne Chase is divided between three counties, Dorset, Wiltshire, and Hampshire. The Ancient Earthworks of the last county are being fully recorded by Dr. J. P. Williams Freeman (to whose friendship and knowledge I am much indebted); accordingly, our surveys have overlapped in this outlying corner of Hampshire. Such overlap, however, could not be avoided in dealing with a district like Cranborne Chase, that is naturally isolated owing to its physical features and surroundings. Its Ancient Earthworks must be studied as a whole, without regard to county boundaries, and the lower valley of the Avon, with the mere, or march, along the western side of the river, mark the natural division between the district now known as Hampshire, and the districts now known as Dorset and Wiltshire.

The method adopted in making this survey has been, first, to make a tracing of the 25 inch Ordnance Survey sheet that delineates the earthwork under examination. Then, to study the 6 inch Ordnance Survey sheet of the same place, in order to note the rise and fall of the land by the contour lines—which are omitted in the 25 inch scale. Then, to

PREFACE

examine the site with both the tracing and the 6 inch sheet, in many cases frequently so as to verify the record and to supplement omissions. And finally, to measure up sections of such banks and ditches belonging to the earthwork under examination, as might seem to be typical.

The original surveys of the enclosure on Rockbourne Down, and of the entrenchment at Horse Down Clump (above Ashcombe), are sketch plans of the respective sites, measured by chain, tape, and rod, but not corrected by instruments: while those of the pastoral enclosures on Woodcuts Common, and on Berwick Down, have been recently made by Mr. Herbert S. Toms, who has kindly supplied me with their plans.

I am conscious that my performance has fallen short of my purpose —but at least I hope that this book of plans presents an intelligible record of the earthworks still remaining within the area undertaken, and further, I hope that it may be of some service to archaeologists; that it may remind the dwellers in and around Cranborne Chase of the ancient relics scattered over our downs; and that such reminder may conduce to the future preservation of these priceless earthen monuments.

H. S.

Cuckoo Hill,
    South Gorley,
        Fordingbridge.

CONTENTS

LIST OF PLANS

HILL-TOP CAMPS

CAMPS ON HIGH GROUND

ENTRENCHED ENCLOSURES, PROBABLY FOR PASTORAL USAGE

CONTENTS

A LIST OF THE ANCIENT EARTHWORKS,

WITH THE ACREAGE AND OWNERSHIP OF THE
VARIOUS CAMPS ON CRANBORNE CHASE

HILL-TOP CAMPS

		ACRES	
1.	Hod Hill	50	Sir Randolph Baker, Bart., M.P.
2.	Hambledon Hill . .	25	Evelyn Clay-Ker-Seymer, Esq.
3.	Castle Ditches (near Tisbury)	23	Lady Arundell.
4.	Badbury Rings . . .	18	The Kingston Lacy Estate.
5.	Whitsbury Castle Ditches	16	The Earl of Shaftesbury.
6.	Winkel-bury	12½	Colonel A. H. Charlesworth.

(Excavated by General Pitt-Rivers in 1881-2, and proved to be probably pre-Roman. See "Excavations in Cranborne Chase," vol. ii.)

		ACRES	
7.	Castle Rings (near Shaftesbury)	11½	George Gordon, Esq.
8.	Buzbury Rings . . .	11	E. G. Farquharson, Esq.
9.	Chiselbury	10	The Earl of Pembroke.
10.	Clearbury Ring . . .	5	The Earl of Radnor.
11.	Damerham Knoll . .	3½	Sir Eyre Coote.
12.	Penbury Knoll . . .	3½	The Earl of Shaftesbury.

CAMPS ON HIGH GROUND

		ACRES	
1.	Bussey Stool Park . .	5½	H. F. W. Farquharson, Esq.
2.	Odstock Copse. (Fragment)	3½	The Earl of Radnor.
3.	Mistleberry Wood . .	2	Captain H. A. Cartwright.
4.	Thickthorn. (Fragment) .	¾	George and Henry Good, Esqrs.
5.	South Lodge, Rushmore .	¾	A. E. Lane-Fox Pitt-Rivers, Esq.

(Thoroughly excavated by General Pitt-Rivers in 1893, and proved to be of the Bronze Age. See "Excavations in Cranborne Chase," vol. iv.)

ENTRENCHED ENCLOSURES, PROBABLY FOR PASTORAL USAGE

ACRES

1. Rockbourne Down . . .	96	The Earl of Shaftesbury.
2. Soldier's Ring	27	Sir Eyre Coote.
3. Chickengrove	12	The Earl of Pembroke.
4. South Tarrant Hinton Down (1)	8	} H. F. W. Farquharson, Esq.
5. South Tarrant Hinton Down (2)	5	
6. Fragment near Vernditch Chase	8	Sir Eyre Coote.
7. Tarrant Hinton Down (near Eastbury)	6	H. F. W. Farquharson, Esq.
8. Near Knighton Hill Buildings.	$2\frac{1}{2}$	The Earl of Pembroke.
9. Martin Down	2	Sir Eyre Coote.

(Thoroughly excavated by General Pitt-Rivers in 1895-96, and proved to be of the Bronze Age. See " Excavations in Cranborne Chase," vol. iv.)

10. Bussey Stool Park . . .	$1\frac{1}{2}$	H. F. W. Farquharson, Esq.
11. Near Woodcuts Common about	2	} A. E. Lane-Fox Pitt-Rivers, Esq.
12. At the South-Eastern end of Woodcuts Common about	2	

(Mr. Herbert S. Toms has recently traced this three-sided enclosure with rounded corners and very slight entrenchment.)

13. Pimperne Down. (Fragment)	—	Viscount Portman.
14. Church Bottom, Prescombe Down	$\frac{3}{4}$	} The Earl of Pembroke.
15. Fifield Down	$\frac{1}{2}$	
16. Handley Hill	$\frac{1}{4}$	A. E. Lane-Fox Pitt-Rivers, Esq.

(Thoroughly excavated by General Pitt-Rivers in 1893, and considered by him to be either of the Bronze Age or Early Roman. See " Excavations in Cranborne Chase," vol. iv.)

17. Near Oakley Down . . .	$\frac{1}{4}$	The Earl of Shaftesbury.
18. Knighton Hill	$\frac{1}{4}$	The Earl of Pembroke.
19. Chettle Down	—	E. W. F. Castleman, Esq.
20. Near Mountslow Cottage .	—	Sir Randolph Baker, Bart., M.P.

A LIST OF THE ANCIENT EARTHWORKS

There are also enclosures that may have been connected with pastoral usage at the British village sites on Berwick Down, and on Horse Down, near More Crichel (fragment).

EARTHWORKS OF EXCEPTIONAL CHARACTER

Knowlton Earthworks . . . . The Earl of Shaftesbury.
Castle Hill, Cranborne . . . . Hector Monro, Esq.
The Mizmaze on Breamore Down . Sir Edward Hulse, Bart.
Castle Green, or Hill, Shaftesbury (almost destroyed).

DYKES AND DITCHES

Bokerly Dyke. (Excavated near Woodyates in 1888-89-90 by General Pitt-Rivers. See " Excavations in Cranborne Chase," vol. iii.) Grim's Ditch. Burcombe Punch Bowl. Row Ditch. Near Crockerton Firs. Half Mile Ditch, on White Sheet Hill. Great Ditch Banks (almost destroyed). South Down, above Eastcombe. Horse Down Clump, above Ashcombe. Hatts Barn. Charlton Down. Melbury Hill. Fontmell Down. Tennerley Ditch (partially destroyed). Launceston Down. Near Mudoak Wood.

BRITISH VILLAGE SITES

1. Woodcuts. 2. Rotherley. 3. Woodyates. (These three sites were thoroughly excavated by General Pitt-Rivers. See " Excavations in Cranborne Chase," vols. i, ii, and iii.) 4. Gussage Down. 5. Blandford Race Down. 6. South Tarrant Hinton Down. 7. Tarrant Hinton Down, near Eastbury. 8. Chettle Down. 9. Near Oakley Lane. 10. Oakley Down. 11. Middle Chase Farm. 12. Near Great Ditch Banks. 13. Marleycombe Hill and Chickengrove. 14. Swallowcliffe Down. 15. Berwick Down. 16. Black-bush on Pentridge. 17. Tidpit Common Down. 18. Horse Down, near More Crichel.

ROMAN WORKS

Ackling Dyke, *i.e.*, the road from Old Sarum to Badbury Rings. The road from Badbury Rings to Ashmore and onward. The road from Badbury Rings towards Hamworthy. Hemsworth Villa. (Excavated by the Rev. G. H. Engleheart. See " Dorset Field Club Proceedings," 1909.)

A LIST OF THE ANCIENT EARTHWORKS

Barton Hill Villa. (Partially excavated in 1845. See Hutchins's "History of Dorset," vol. i, p. 319, edition of 1861.) Villa near Iwerne Minster. (The last excavation conducted by General Pitt-Rivers, and not recorded.) Lydsbury Rings, on Hod Hill (?).

Besides the above Earthworks there are a very large number of Barrows scattered over this district. The Map of Cranborne Chase will show their general disposition, but the round Barrows of the Bronze Age are so numerous here, and so often partly destroyed by cultivation, that I fear omissions may be found. The following list records the comparatively rare Long Barrows of the Stone Age:

1. Pimperne Long Barrow.
2. Chettle Long Barrow.
3. Blandford Race Down (near Vanity Hill Wood).
4. Little Down (near Langton Lodge).
5. Thickthorn Down.
6, 7. Gussage Down.
8, 9, 10. On the western side of Bokerly Dyke. (These are unusual, having no side ditches.)
11. Round Clump, near Great Yews.
12. Knap Barrow, Knoll Hill.
13. Hambledon Hill. (These two last are unusual, having no side ditches.)

The axes of all these Barrows point N.W. and S.E. approximately.

14. White Sheet Hill, near Half Mile Ditch.
15. Near Water Lake, in the line of "The Cursus."
16. Giant's Grave, Breamore Down.

The axes of these Barrows point N.E. and S.W. approximately.

17. Chettle, near Tarrant Hinton Down.
18. Telegraph Clump, Blandford Race Down.
19. Near Tidpit Common Down (much wasted).

The axes of these Barrows point E. and W. approximately.

20. Worbarrow. (Excavated by General Pitt-Rivers. See "Excavations in Cranborne Chase," vol. iv.)[1]
21. Giant's Grave, near Wick.
22. Duck's Nest, Rockbourne Down.
23. Grans Barrow, Knoll Hill. (Unusual, having no side ditches.)

The axes of these Barrows point N. and S. approximately.

[1] Models of all these sites, excavated by General Pitt-Rivers, may be seen in his museum near Farnham, Dorset.

A LIST OF THE ANCIENT EARTHWORKS

Cultivation Banks, or Linchets, that appear to be of great antiquity,[1] may be found in especial number near the following places:

On the Western side of Melbury Hill, near the Shaftesbury and Blandford road.

On the Southern and Western sides of Melbury Down.

Near Fontmell Magna.

Around Rushmore.

On the Southern side of Blandford Race Down.

On Chettle Down.

On Bottlebush Down.

On the Southern and Western sides of Pentridge.

On the Down towards Toyd Farm, North of Grans Barrow and Knap Barrow.

On the Eastern side of the Combe down which runs the Blandford and Salisbury road into Coombe Bissett.

On the sides of the Combes which run up into the chalk ridges, North and South of Broad Chalke.

Near Knighton Hill Buildings.

Near Little Yews, and Clearbury Ring.

[1] See "The Village Community," by G. L. Gomme, and "Folk Memory," by Walter Johnson.

'TIS time to observe Occurrences, and let nothing remarkable escape us; The Supinity of elder dayes hath left so much in silence, or time hath so martyred the Records, that the most industrious heads do finde no easie work to erect a new *Britannia*. 'Tis opportune to look back upon old times, and contemplate our Forefathers.—SIR THOMAS BROWNE.

INTRODUCTION

THE ruined Buildings of Ancient History either stand near present habitation, or in situations that still appeal to us as being desirable. Roman walls and gateways survive within our towns; Mediaeval Castles dominate modern streets; and Monastic Buildings stand in sheltered valleys bearing witness to a choice that we still endorse. But when we gaze farther back, farther up the stream of time, and when we seek for the relics left by prehistoric men, we find ourselves in places where solitude now reigns. Their camps, their settlements, their cultivation banks, and their boundary ditches were on the hills, remote from present habitation, for the sites which they sought have long since been abandoned. Life has receded from the hill-tops.

Once upon a time great white banks and mounds of upturned chalk crowned our hills, proclaiming the camps of tribal safety, and the tombs of the mighty dead—while lesser white banks rambled up and down the open country, and steep scarps lined the hill-sides, proclaiming pastoral boundaries and primitive cultivation. Then, safety lay on the hill-top. Then, the valleys were shunned, probably because they were swamps and wolf-haunted. But we have abandoned the choice of prehistoric men. Now, down in the fertile valley, shrouded in smoke, lies the modern town, while the rush of the train, and the hoot of the motor tell where our traffic passes. No one seeks the old sites nowadays—except shepherds and their flocks, or sportsmen, or pilgrims in search of the past.

Ichabod may seem to be written on ruined buildings, for they testify to a glory that is departed, but the grassy mounds, and ditches, and hollows of prehistoric sites strike a deeper note of desolation. Here the labours of our forefathers have reverted to Mother Earth, to wild nature, and to the elemental ministry of the seasons.

It is difficult to express the genius of such places, or the dim sense of communication and realization which they impart. Mere proximity, contact, counts for something, however, although it may elude words. The following story expresses this ideal claim. Robert Browning was in

Paris with his son, then a little boy, and in some public place pointed out an old gentleman sitting on a seat—"Go up and touch him"—the boy obeyed; when he came back his father said, "Now, when you are a man, remember that to-day you have touched Béranger"—Virtue may go out even to those who only touch the hem of a garment. So Places as well as Persons may impart something to your touch, and it is this reverent touch, inspired by admiration, that I wish to bring to these derelict places that once were inhabited by prehistoric men. This is my purpose. I want to take my reader a far journey across the chalk uplands of Cranborne Chase, in which we may revive the wonder of this primitive life—so remote from our own; and in which we may widen our outlook on familiar scenes as we seek the origins of our present landmarks.

In these days it is possible to look far afield, and to know wide stretches of country, in a manner that was almost impossible to our more rooted forefathers. Formerly, only the horseman could go where we are going; but a horse is a tiresome servant while you wait, and must always be the first care of his rider. We shall do better to depend on our own legs, and on our own seven league boots. What bliss it was in child-hood to read of the magical seven league boots, and to see their possessor, in Cruikshank's etching, striding across a landscape! What a gift of the gods, such power of locomotion! What a fantastical conception! And lo, since our childhood, the fairy tale has become fact—common fact so far as concerns our power over seven leagues; for *wheels* were meant though *boots* were written; far and near have changed their relative meaning; every blacksmith in the remotest hamlet can restore our stride if mischance befalls, and so, on silent wheels we can now range far and wide across the country. A bicycle is our magic of seven leagues. It carries us where we will, and then, laid on the grass, it waits our fur-ther will. A few drops of oil and our own muscles are its sole demand, while in return it endows us with the magical power of the old fairy tale.

Thus equipped, let us range the chalk summits of Cranborne Chase in quest of earthworks, and our wandering survey will show where lay the prehistoric sites of safety, and the pastoral lands of remote antiquity, and the origins of cultivation. From these we may learn something of the compelling circumstances that shaped life in this unrecorded period, and of the meaning of the humps, and hollows, and dykes, and ditches upon the downs, and we shall gain a new respect for these forsaken earthworks that were wrought by the flint tools, the horn picks, and the patient cunning of our forefathers.

INTRODUCTION

At the outset I would offer a tribute of admiration to the great work done by General Pitt-Rivers, and to its record contained in the four volumes of his " Excavations in Cranborne Chase." He set a standard that is pre-eminent in its many-sided excellence, and he brought to bear upon his subject a wide experience of primitive antiquities, a genius for the study of origins, and an understanding of the nature of things, that resulted in these invaluable works. They are monuments of original research, and of exact record.

Earthworks can only be really understood by spade-work. The finds may be few, and of a cast-away description, but notwithstanding, they are positive in their evidence, and it is from excavations in the first place that we can increase our knowledge of prehistoric life. We all desire wishing caps, and if mine is ever fitted, I shall wish for another archaeological landowner who shall dig for further knowledge on Cranborne Chase. There is so much to be done. So many questions that are suggested by a superficial survey, and that only excavation can answer. However, excavation has to be regarded as a counsel of perfection, for it cannot cover the field—the earthworks are too numerous—and thus a survey of the varied earthworks on Cranborne Chase may be of help in their interpretation, and of value as a record of landmarks that are always liable to destruction. For destruction is what happens under cultivation. Ancient banks are spread and ditches filled in, till gradually a low rise and fall in the land is all that remains as the present witness of a past site. In some cases the obliteration is complete. Everywhere the plough has been at work destroying these earthen marks of pre-historic men. The Farmer cannot be expected to pay rent on behalf of the Archaeologist, and entrenched sites have generally been regarded merely as humpy ground. When Dr. Stukeley, in 1723, explored the Roman Road from Old Sarum to Winchester, he chanced upon a sample of this destruction—"This way paſſes the river Bourn at Ford: the ridge of it is plain, though the countrymen has attacked it vigorouſly on both ſides with their ploughs: we caught them at their ſacrilegious work, and reprehended them for it." Poor countrymen, scolded by archaeologists, ignored by legislators, but well remembered by the tax-collector. The Archaeologist who catches them at their sacrilegious work, and repre-hends them for it, forgets the rent, the risks, and the difficult livelihood of the countryman. The burden of the preservation of ancient earthworks needs to be placed on broader shoulders than those that have hitherto chanced to bear them. Really, we owe some gratitude to the peaceful cultivator for the fragments that remain.

Still, the loss is deplorable that we have suffered through these vigorous

3

attacks of the countryman. This loss has come from lack of national imagination. If we had realized the meaning of these useless earthworks they would have been preserved, with reservations in farm leases, and exemption from all incidence of taxation; and so when present life forgets past life it is well to remind, and to draw attention to the ancient places that were chosen and entrenched by our forefathers.

It is curious that the old cartographers, Saxton, Norden, and Speed, did not mark camps and earthworks in their surveys. Speed records a few in his letterpress descriptions of the counties, and in writing of Dorset he mentions Maumbury, Poundbury, Maiden Castle, and Badbury—but that is all. Evidently they were held in small estimation, a neglect that increases the debt of gratitude that we owe to Dr. Stukeley, whose "Itinerarium Curiosum" (published in 1724, 2nd edition 1776) was the first contribution to a study by means of plans of these relics of our History.

When maps were few, and surveys scant, how exciting must have been the search for primitive earthworks! Imagine a description of Dorset—as Speed describes it—with never a word about the camps on Hod Hill or on Hambledon Hill, and with no mention of Bokerly Dyke! And then think of riding afield as a roving enquirer, and coming upon these forgotten earthworks that express such indomitable energy, and that confront us with such great problems of prehistoric life. This was the happy fortune of the archaeologist in the eighteenth century.

What Dr. Stukeley began, Sir Richard Colt Hoare continued. In the early years of the nineteenth century Sir Richard Colt Hoare gave up hunting foxes in favour of hunting earthworks, and the ardour of his new chase led him across the borders of Wiltshire, into Dorset, and the district of our survey. His folio volumes on "Ancient Wiltshire" contain most interesting plans of some of the earthworks on Cranborne Chase, surveyed by Crockett and engraved by Basire, and these were the best plans that had been done of ancient sites in this part of England. Certainly in beauty of engraving they could not be surpassed; in accuracy and understanding, however, they are in some instances found wanting.

Mr. Charles Warne's map of "Dorsetshire: Its Vestiges, Celtic, Roman, Saxon, and Danish" includes some of the earthworks on Cranborne Chase, but this beautiful map merely locates sites, and moreover it is too small to give any details of plan.

In "Ancient Dorset," by the same author, there are some wood engravings of camp plans, but they are quite unworthy of their place in his book. The wood-engraver must have done his work without the personal supervision of the author. The most accurate plans of earthworks on Cranborne Chase—as might be supposed—are to be found in General Pitt-Rivers' "Excavations in Cranborne Chase," but these are limited to the earthworks which he excavated. The map of "Ancient Dorset, Wilts, Somerset, and part of Hants," contained in his third volume, only indicates the relative positions of the principal sites. Accordingly, if we wish to study plans of the earthworks on Cranborne Chase we must obtain the 6 inch and the 25 inch Ordnance Survey sheets that set forth this district. The Ordnance Survey is a most admirable and exact work, from its own point of view, but it must not be cited as the final Court of Appeal on matters of archaeology. There are omissions and there are misunderstandings in its record; accordingly, such a survey as I am here attempting may help to perfect the mapped record of these earthworks, that hold their secrets, notwithstanding, for the excavator.

The reader of the subsequent plans will assuredly want to know when these various earthworks were originally thrown up, and, I fear, will not find many definite replies to such enquiry in my descriptions. Without excavation it is impossible to be sure, and, except in a few instances, the great entrenchments on Cranborne Chase have not been excavated with the purpose of attempting to discover their period by means of sections cut through their ramparts and ditches. We can say, however, from such evidence as we now possess that our camps are pre-Roman—probably of the Bronze Age (ending *circa* 500 B.C.), possibly in some cases of the Neolithic Age (ending *circa* 1800 B.C.), but there is much spade-work to be done before we can assign distinctive periods, without qualification, to the great earthworks that are delineated in this book of plans.

Cranborne Chase is a peculiar district. It lies apart from railroads, and apart from most of the road traffic that passes through Ringwood or Salisbury. It is a solitary tract of down-land, corn-land, wood-land, and waste. Dry valleys run far up into the steep flanks of the chalk ridge that is the backbone of the Chase. Streams emerge with intermittent flow in the lower slopes of these valleys. With the exception of Ashmore, Shaftesbury, and Whitsbury, which are set on hill-tops, the villages are scattered in the lowlands. Barrows, long and round: Camps of defence and of safety: Boundary banks and ditches: Pastoral enclosures: Cultivation banks: Roman roads: the sites of many British villages on the uplands, and Dykes of defence, all testify to the former habitation and desirability of this now solitary land. The evidence of its earthworks points to the assumption that a greater population was once settled on Cranborne Chase than is settled there now, and that it was under some sort of cultivation from prehistoric times.

We may gain an idea of the value possessed by this tract of country by considering its natural conditions.

On the East it was bounded by the New Forest. On the South by Holt Forest, and the heathland of Dorset. On the West by the Forest of Blackmore, and on the North by woodlands that probably covered the rich green-sand of the valley of the Nadder—Wastes and Forests that impeded the primitive cultivator, either from the poverty of their soil, or from the tangle of their growth. Amid such surroundings, the rolling chalk hills of Cranborne Chase must have emerged as a desirable land. The soil must have been the same then as now. A retentive loamy chalk ranging to a poor chalk, with clay-capped hills. The water supply must have been better than it is now. The rainfall we believe to have been greater, and the water-level we know, from the evidence of General Pitt-Rivers' excavations at Woodcuts, was higher then, than now. Think of the Tarrant, the Allen, the long Crichel, and the Gussage brooks, the Crane, the Martin Allen, the Rockbourne brook, the Ebble, the Donhead brook, the Iwerne brook, and the Pimperne brook. Think of all these streams flowing constantly fifty feet above their present rise, and you get a very different conception of the prehistoric pastoral and agricultural value of this tract of land. Looking back 2,000 years, we may imagine the area now known as Cranborne Chase as a fairly well watered downland, intersected by scattered woodland that stretched from Blagdon Hill to Holt, from Vernditch to Rushmore, and from Chettered to Ashmore. A truly desirable land when contrasted with its surroundings.

Such were the natural conditions of this tract of country which was chosen by the primitive herdsman and the subsequent cultivator, and the pastoral enclosures and cultivation banks are evidences of a tolerable security for settled possession. They imply that prehistoric men counted on folding their flocks and cattle on the same sites year in and year out, and on gaining the distant reward of laborious cultivation—on reaping where they had sown.

The Hill Camps, with their eminent positions, their great banks and deep ditches, and their concentration, appeal more to our imagination than do earthen folds and lynchets. But the latter were the outcome of the former. This land was desirable for pastoral and agricultural purposes; it was worth defending; and the ancient earthworks on Cranborne Chase proclaim the herdsman and the peaceful cultivator as much as the warrior.

INTRODUCTION

The Roman occupation must have increased the habitable convenience of this district; for the Roman genius created communication with the outer world, and along the highways from Hamworthy and Dorchester came merchandise that furnished the demands of the Romano-British civilization. The excavations made by General Pitt-Rivers in the settlements of Woodcuts and of Woodyates, revealed accessories of life that suggest comfort, *e.g.*, timber and plaster houses, with roofs of shale or tile, hypocausts, furniture, metal-work, fine Samian pottery, glass, jewellery, oysters, etc. We may regard this district at the time of the Roman evacuation as having been thickly populated, peacefully occupied, and well cultivated, and we may attribute the stubborn resistance that later on was here offered to the oncoming West Saxon, as a testimony to the value of the territory which the Romanized Britons were defending.

The limits of this district of Cranborne Chase have been the cause of much contention. But the contention arose respecting the rights attached to the Inner Bounds, and the Outer Bounds of the Chase. With these rights, and their attendant wrongs, we have no concern here. The Outer Bounds, or Limits of the Chase, are the only detail which now concerns us, for these have been taken as the limits of this survey. They are founded, as I have said, upon natural features, that have always tended to impart a certain local and separate character to this district. Accordingly, let us turn to the mediaeval record of the delimitation of Cranborne Chase, within which Perambulation our survey of ancient earthworks is to be confined. "A Chronicle of Cranborne," by Dr. Wake Smart, contains a reproduction of a *Map of Cranborne Chase, from an original Map taken and drawn by Richard Hardinge of Blandford,* A.D. 1618, *which was copied by Mathew Hardinge of Blandford,* A.D. 1677, *and diligently examined and compared with the original by Hy. Dolling.* And "King John's House," by General Pitt-Rivers, contains a reproduction of *A Mappe or Plott of Cranburne Chace Lying in Dorſetſhire, Wiltſhire, and part of Hampſhire, wherein two sortes of Bounds of the said chace are sett forth and expreſſed called the Large Boundes and the Short Bounds. By vertue of a Commiſſion out of his Highnes Court of Exchecquer Directed to Sr. Francis Popham, Sr. John Dauncy, Sr. Antho. Hungerford Knightes, and Thomas Hinton Eſquier. Plotted and performed by Thomas Aldwell, And others. Anno Dñi* 1618. These two maps, and the Perambulations recorded 8 Edward I and 29 Henry III, have been taken as the authorities for the outer Bounds of the Chase; and these outer Bounds have been taken as the circle that limits this survey, because, as previously noted, they fairly represent the primitive natural boundaries that surrounded this district, that obstructed the Saxon advance westward for more than half a century (520 to 577), and that partially exist even to this day. Within these limits, as shown in the general map at the beginning of this book, we shall now proceed to investigate the ancient earthworks on Cranborne Chase.

TO THE READER

AT the outset it may be well to explain the conventions that are used to express the ground plans of earthworks.

The thick ends of the shading lines, or *hachures,* denote the top of the banks, the thin ends, the bottom, while ditches or sunken areas are denoted by dots.

The diagram below shows a section of a ditch between two banks, and its equivalent linear expressions as a ground plan.

Section.

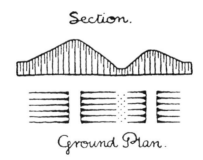

Ground Plan.

I AM told there are people who do not care for maps, and find it hard to believe. The names, the shapes of woodlands, the courses of the roads and rivers, the prehistoric footsteps of man still distinctly traceable up hill and down dale, the mills and the ruins, the ponds and the ferries, perhaps the *Standing Stone* or the *Druidic Circle* on the heath; here is an inexhaustible fund of interest for any man with eyes to see or twopence-worth of imagination to understand with.—R. L. STEVENSON.

HILL-TOP CAMPS

THE CAMPS ON HOD HILL, AND ON HAMBLEDON HILL

HOD HILL and Hambledon Hill are the two summits of a chalk outlier from the main ridge of the Dorset Northern Downs.

The Stour separates them on the West from Shillingstone Hill, and the Iwerne brook on the East from the Cranborne Chase uplands, while between them there is a dry valley cleft, which is crossed by the road from Steepleton to Hanford. Hambledon Hill rises to a height of 623 feet at the Southern end of the camp, and its ridge juts out for a mile into Blackmore Vale, as a narrow bare promontory with steep scarps on the North, the East, and the West.

Hod Hill is less eminent. It is 471 feet in height at the North-Western corner of the camp area—within the low lines of the Roman entrenchment. The slopes of the hill are gradual on the North, the East, and the South, but on the West the hill scarp mounts abruptly for 300 feet above the Stour with a rise of 1 foot in 2 feet.[1]

Whatever doubtful problems these two camps may suggest, we shall not doubt their makers' choice. These hills are natural strongholds, made stronger by man's indomitable energy and skilful purpose. Their castrametation compares with that of Castle Ditches, near Tisbury, of Whitsbury Castle Ditches, and of Badbury Rings; but the actual camps are larger than any of these—the area of Hod Hill being about 50 acres, and of Hambledon Hill about 25 acres.

HOD HILL

THE camp on Hod Hill was surrounded on the North, the East, and the South sides by triple entrenchments, and on the West by double entrenchments along the steep scarp that falls abruptly to the Stour

---

[1] This remarkable cleavage made by the Stour through the chalk ridge of North Dorset is discussed and explained by A. J. Jukes Brown, " Dorset Field Club Proceedings," xvi, 1895.

Valley. The outer bank and ditch on the East and South sides have been partially effaced by cultivation, but they can be traced all round, with gaps in their continuity. The Ordnance Survey ignores this continuity, hence the frequent description of Hod Hill camp as being surrounded by double entrenchments.

At the South-Western corner, the entrance from the Stour Valley is commanded by an outer, flanking bastion, and here the defences are specially strong. The inner bank rises to the prodigious height of 41 feet above the ditch, with a rise of 1 foot in 2 feet. To scramble up such a bank with a measuring rod is not easy; imagine such a scramble with a fierce stone-throwing Briton above! We get some idea of the defences of these camps even by peaceful survey.

This may be supposed to have been the most important entrance, as it would always have been needed for the defenders of the camp in order to obtain access to water—the Stour. There is another specially defended entrance on the Eastern side above Steepleton. Here the approach to the camp winds between ramparts that must have commanded an enemy on their left flank as they struggled up the narrow pathway of danger; while the incurving horns of the inner bank gave further protection to the defenders of the area. Besides these there are three other entrances which may, or may not have been original.

At the North-Western corner of the camp area there is a low, precisely cut entrenchment that has generally been assumed to be of Roman origin. If the plan and sections are studied, the absolute difference of this inner entrenchment from the outer entrenchments will be recognized. From superficial survey I see no reason to suppose this earthwork to be other than Roman. Its low elevation implies defenders that depended upon discipline rather than upon a big bank for protection. And the precise execution is wholly different from any of the earthworks on Cranborne Chase, Soldier's Ring excepted. The neat finish of this entrenchment should be noted where it fits on to the Western scarp of the great camp rampart (see plan), and the returns of the ditches on either side of the two entrances; also the extra bank at the corner, which would be the weakest point of defence.

Hutchins's " History of Dorset," vol. i, p. 306, edition of 1861, gives a " Plan of British and Roman entrenchments on Hod Hill," and the following extract gives a description of the area of the Roman camp, since ploughed up. " The whole appears to be formed with the greatest " regularity and precision, and the same order seems to have marked the " disposition of the interior. The marks of tents or huts may still be traced " at regular intervals, and appear to have been placed in lines facing the

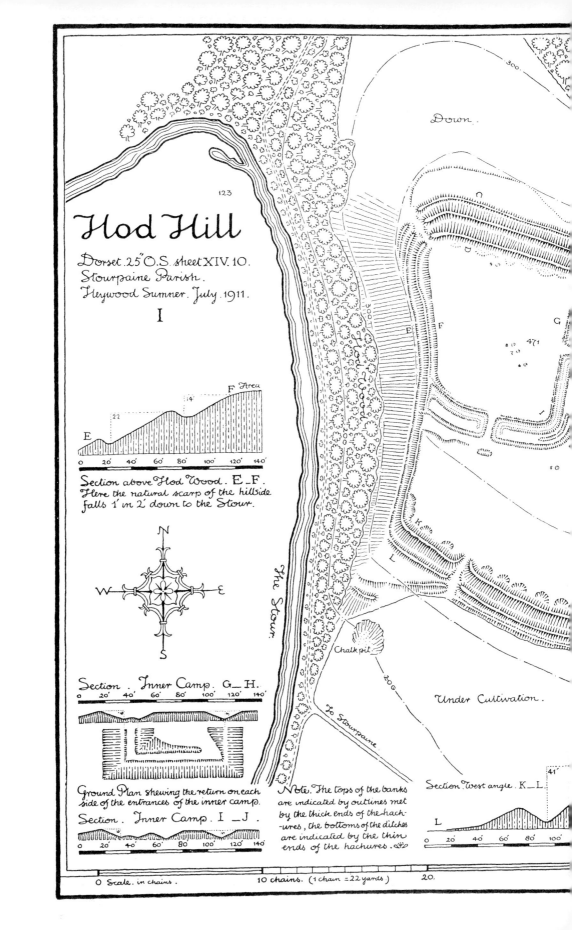

123

Hod Hill

Dorset. 25″ O.S. sheet XIV. 10.
Stourpaine Parish.
Heywood Sumner. July. 1911.

I

Section above Hod Wood. E_F.
Here the natural scarp of the hillside
falls 1′ in 2′ down to the Stour.

N
W · E
S

Section . Inner Camp. G_H.

Ground Plan shewing the return on each
side of the entrances of the inner camp.

Section . Inner Camp. I _J.

Note. The tops of the banks
are indicated by outlines met
by the thick ends of the hachures, the bottoms of the ditches
are indicated by the thin
ends of the hachures.

Down.

Hod Wood

471

E F G

K
L

Chalk pit

To Stourpaine

Under Cultivation.

Section West angle. K_L.
41

The Stour.

0 Scale. in chains.   10 chains. (1 chain = 22 yards)   20.

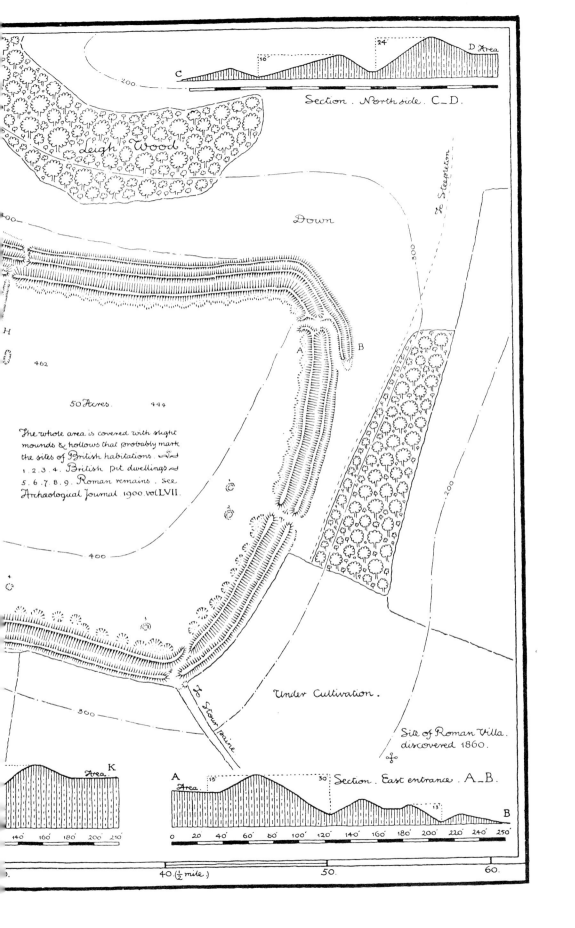

Section. North side. C_D.

Leigh Wood

Down

Steepleton

H

462

50 Acres.   444

The whole area is covered with slight
mounds & hollows that probably mark
the sites of British habitations.
1.2.3.4. British pit dwellings
5.6.7.8.9. Roman remains. See
Archæological Journal 1900. vol LVII.

Under Cultivation.

Site of Roman Villa
discovered 1860.

Area K

Area

Section. East entrance. A_B.

140' 160' 180' 200' 210'

0 20' 40' 60' 80' 100' 120' 140' 160' 180' 200' 220' 240' 250'

40. (½ mile.)   50.   60.

" front of the camp, three or four deep, with a large open space between
" them and the entrenchments. Wide level roads intersected the camp
" from each entrance. There can be but little question as to the origin of
" this work; every surviving portion answers perfectly to the system of
" encampment followed by the Romans, and so minutely described by
" Polybius (bearing in mind, of course, the difference rendered necessary
" by the smaller size of this work), and the imagination is wonderfully
" assisted by the configuration of the surface in supplying doubtful links.
" This camp, therefore, acquires extraordinary interest, if we call to mind
" that in all probability it is one of the most perfect examples known of the
" Roman entrenched camp."

After reading the foregoing description of what was to be seen
within the area of this small Roman camp, it is very disappointing to
turn to the plan that was taken in 1858 (p. 307). The area is a mere
blank! Hutchins gives some account of numerous relics that have been
dug up from time to time within the area of the camp on Hod Hill.
They were in Mr. Durden's collection at Blandford, but are now in the
British Museum. Roach Smith ("Collectanea Antiqua," vol. vi) says of
them: "These antiquities have been collected by Mr. Durden during a
" considerable number of years in the course of agricultural operations."
From which it does not appear that there had been systematic excavation
at Hod Hill. The iron weapons found point to occupation during the
later portion of the prehistoric Iron age. The Roman camp, or Lydsbury
Rings as it is called, yielded a number of Roman relics, turned up
apparently by the plough. The coins found give a very early date to
the Roman occupation of this portion of Hod Hill.

The disappointing plan given in Hutchins's "History of Dorset,"
was "taken in 1858, just as the workmen were paring the turf preparatory
" to cropping the western portion of the area, and prior to the ancient
" traces being obliterated." Lydsbury Rings had not then been disturbed,
but its area was subsequently ploughed over. In the preface to "Dorsetshire: Its Vestiges, Celtic, Roman, Saxon, and Danish," by Charles
Warne, 1865, he says: "Thus, and that very recently, the Roman
" Castrum within the Celtic camp and Oppidum on Hod Hill has, to the
" lasting disgrace of those concerned, been ruthlessly destroyed; the plough
" has passed over its Praetorium, and the site once occupied by the surround-
" ing host, with all its details so well defined, is now no longer to be traced;
" thus an example of Roman castrametation the finest of its kind, in fact
" unique, has been obliterated, and that without a voice being raised or an
" effort made to stay the hand of the despoilers."

Accordingly, from these two extracts we may assume that the whole of the areas of the outer and the inner camps on Hod Hill have been ploughed over. Now they have reverted to down grass, and we have to be thankful that the plough spared the low banks and shallow ditches that still mark the defences of Lydsbury Rings.

In the autumn of 1899 excavations were carried out within the area of Hod Hill Camp by Sir Talbot Baker, under the supervision of Prof. Boyd Dawkins, and the results are recorded in the " Archaeological Journal," 1900, vol. lvii.[1] Circular hut enclosures and pits were the earthworks that were examined, and from the relics found in them, " it " may be inferred that the settlement on Hod Hill continued to exist from " the pre-Roman age well into the time when the Roman influence was " dominant in this district." No sections were cut through the defensive banks and ditches, consequently there is no positive evidence as to the original construction or as to the date, either of the outer defences of Hod Hill Camp, or of the inner defences of Lydsbury Rings. In " Earthwork of England," by Hadrian Allcroft, Hod Hill and its earthworks are fully discussed, pp. 361-367. Mr. Hadrian Allcroft does not accept the record in Hutchins which I have quoted, nor the (rather vague) evidence of Mr. Durden's finds, but considers that Lydsbury Rings may have been thrown up by the Romano-British in opposition to the West Saxon advance, and cites " the lofty irregular vallum and fosse, the object of " which is not obvious," that cut across the adjoining camp on Hambledon Hill, as a parallel instance of a lesser camp being made within a larger one. I do not think, however, that the two earthworks compare. The one low, calculated, and very precisely cut, the other lofty, haphazard, and rudely thrown up. I take Bokerly Dyke as the proved type of earthwork that the Romanized Britons threw up to oppose the West Saxon advance. The mind that planned Lydsbury Rings, and the method of defence implied by their castrametation appear to be wholly different from the mind and purpose behind the earthworks within the area of Hambledon Hill Camp.

The inner rampart that defends the great circuit of the camp on Hod Hill seems to have been strengthened—raised—on three sides, North, East, and South, by means of excavations from the area side. In places these take the shape of semicircular hollows from which spring the ramparts, in other places the shape of a broad and shallow ditch. The summit also of this inner rampart is humpy and irregular, while the

---

[1] In this record the area of Hod Hill is given as "about 320 acres," and of Lydsbury Rings as "about 70 acres." Although this is an obvious error, it may be well to note that the whole area of Hod Hill Camp is about 50 acres, and the area of the Northern corner enclosed by Lydsbury Rings is about 7 acres.

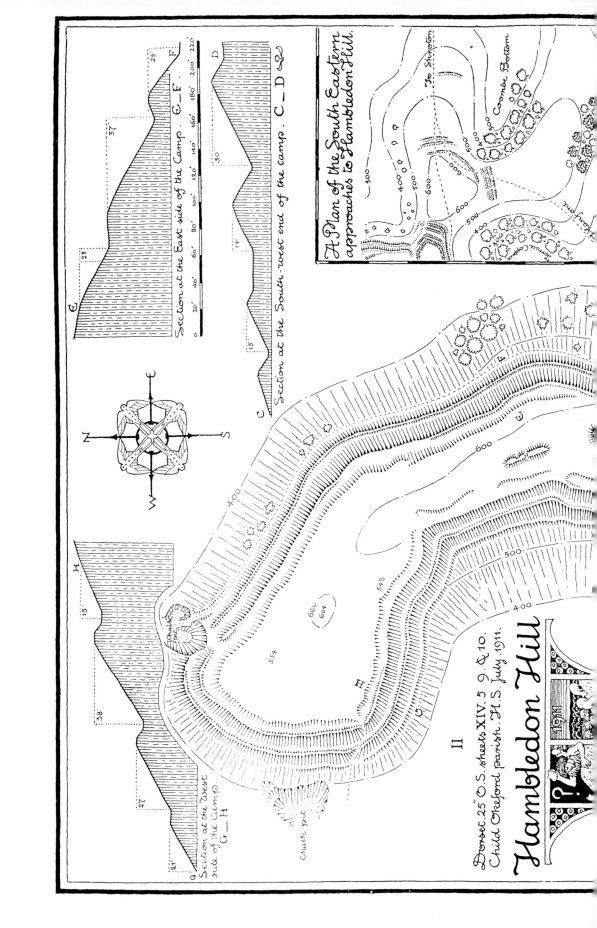

A Plan of the South Eastern
approaches to Hambledon Hill.

To Shroton

Coombe Bottom

Hanford

Section at the East side of the Camp. G_F.

Section at the South-west end of the camp. C_D.

Section at the West
side of the Camp.
G_H

Hambledon Hill

Dorset 25" O.S. sheets XIV. 5. 9. & 10.
Child Okeford parish. H.S. July 1911.

II

1911

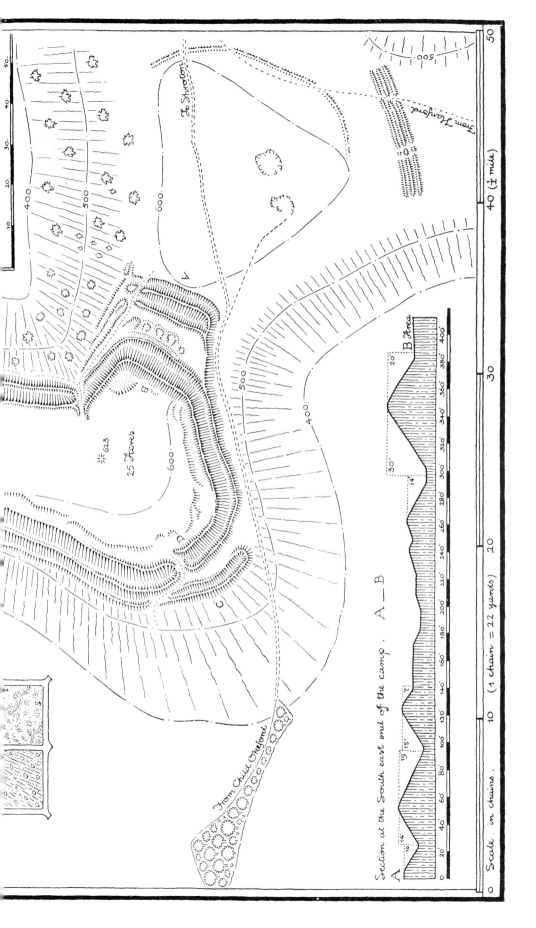

Section at the South east end of the camp. A — B

Scale in chains. (1 chain = 22 yards)

To Shroton

From Hanford

From Child Okeford

25 Acres.

623

A Area

B Area

summits of the outer ramparts are even and continuous. This may be well seen from below, looking up at the camp from the Steepleton and Hanford road. The present appearance of the inner rampart suggests sudden emergency, and hasty reinforcement of a bank that was lower, and contrasts with the even run of the outer ramparts and ditches, and with the orderly embanking of the Eastern and South-Western entrances. A section cut through the triple entrenchments of Hod Hill Camp, and another cut through the low defences of Lydsbury Rings, would probably give us a final answer to the conjectures that have been put forward respecting these earthworks.[1]

The spacious area of this camp is covered with low humps and shallow hollows—now too much wasted by cultivation for definite survey record. They suggest that this site was sought after, and fully occupied in prehistoric times.

The excavations by Professor Boyd Dawkins, to which I have referred, showed that Roman relics were only found within and near Lydsbury Rings. Apparently but a small portion of the area was occupied by the Romans. And the coins found suggest that their stay was short. When the military necessity of occupying this outpost ceased, the site seems to have been abandoned. There are no signs of continuous occupation throughout the Roman period, such as General Pitt-Rivers found by his excavations at Woodcuts, Rotherley, and Woodyates.

HAMBLEDON HILL

THE natural approach to Hambledon Camp is on the South-Eastern side. Here the hill ridgeway forks into two spurs, one trending East towards Shroton, the other South-East towards the pass road leading from Steepleton to Hanford. Both these spurs are crossed from scarp to scarp by low banks and shallow ditches, double and triple, of which there are so many examples on Cranborne Chase.

The down outside the South-Eastern defences of Hambledon Hill Camp has been dinted with modern diggings for flints, and thus it is impossible to form an opinion as to what sort of habitation existed here to account for these outlying banks and ditches. But we may be fairly sure that these simple multiplications of low banks and shallow ditches

---

[1] Mr. Reginald A. Smith in the " British Museum Guide to the Antiquities of the Early Iron Age," p. 122, says: " Hod Hill is one of the most imposing Dorset heights that were " crowned with earthworks for the protection of the inhabitants during the Bronze and, possibly, " in some cases, the Neolithic period."

belong to a different period to that of the great earthworks, so cunningly planned, that encircle and defend the approaches of the Camp on Hambledon Hill.

The South-Eastern defences of the Camp are very remarkable. For miles around the great inner bank is a landmark. It rises 30 feet above the bottom of the ditch, and 24 feet above the area, and beyond the double bank and ditch of this formidable earthwork there is a broad berm of 100 feet, very uneven in surface and protected at its South-Eastern extremity by two more great banks and ditches. The entrance passes through this outwork close to the Northern scarp of the ridge, and in such a way (see plan) that an enemy would be assailed for 200 yards on the flank by the defenders of the camp. The random digging of the berm between the inner camp lines and the outer defences is noticeable, and the sudden rise of the inner bank as it crosses the down ridge, with the rough scoops into the area—whence presumably came the earth—suggest emergency, and strengthening of existing defences. The Western entrance is also very strongly defended (see plan). In both cases these earthworks are in a fine state of preservation.

The Northern entrance of the Camp does not show signs of much usage. It must always have been inconvenient owing to its precipitous gradient. The South-Eastern and South-Western entrances were apparently the usual approaches, and on these the Camp defenders expended their utmost skill in fortification.

Within the Camp area—which at the Northern end rises nearly 100 feet above the triple entrenchments—there are some curious earthworks of debatable purpose, and a round barrow.

The long mound Northward of the said debatable earthworks is a doubtful long barrow. It may compare with the long mound within the area of Knap Hill Camp, excavated by Mr. and Mrs. Cunnington ("Wilts Archaeological Magazine," vol. xxxvii, p. 42), and shown by the finds to have been thrown up at some time during or after the Roman period—probably as a shelter.

Shelter is a requirement that is forced on the attention of any modern excavator on these uplands. The wind sometimes sweeps and buffets across these bare downs with such rigour that it becomes almost unbearable; but as the digging proceeds, so shelter is obtained, both from the bank of upturned soil and from the excavated and lowered ground level. Thus protected, it is possible to dig in peace while the wind whistles through the bent grass above. Cattle need shelter as much as men; and it seems possible that the curious horseshoe form of bank, enclosing a sunken area, that may be found here, on Hambledon Hill, on

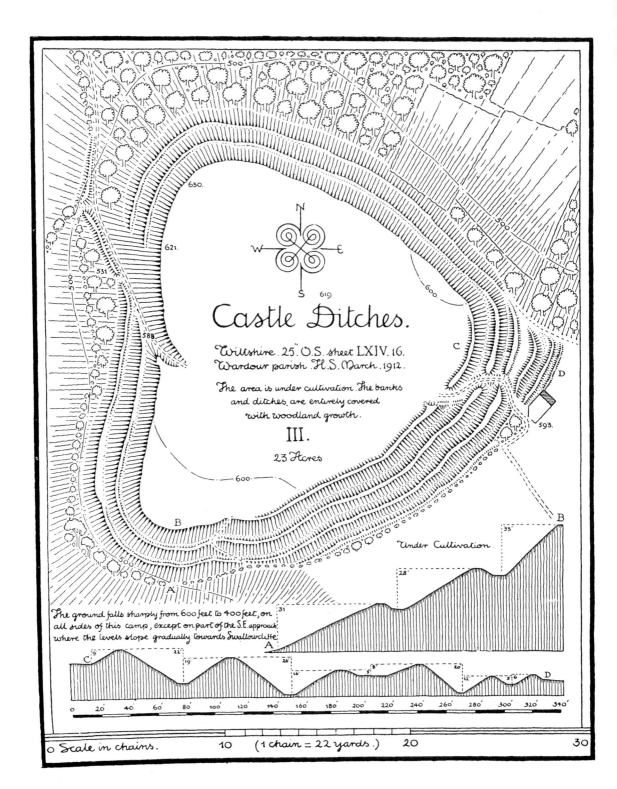

Castle Ditches.

Wiltshire. 25". O.S. sheet LXIV. 16.
Wardour parish. H.S. March. 1912.

The area is under cultivation. The banks
and ditches, are entirely covered
with woodland growth.

III.

23 Acres

Under Cultivation

The ground falls sharply from 600 feet to 400 feet, on
all sides of this camp, except on part of the S.E. approach,
where the levels slope gradually towards Swallowcliffe.

o Scale in chains.   10   (1 chain = 22 yards.)   20   30

CASTLE DITCHES

Gussage Down, on Blandford Race-Down, on the Tarrant Hinton Downs, on Chettle Down, and at Buzbury Rings, may have been made for this elemental purpose.

It is impossible to write of Hod Hill and of Hambledon Hill and merely to expatiate on their prehistoric camps. Their names will always raise up visions of beauty, the memory of which abide as a possession to those who know these mighty earthworks. Hod Hill, with its bald, mound-like summit, here and there fringed with beech woods, dominating the Stour Valley, and rising abruptly, like a rampart above the still reaches of the winding river—Hambledon Hill, with its down scarps spotted with yews and thorn trees, with thickets of ash, elder, white beam, and yew, over which great wisps of Traveller's Joy fling their feathery tangle, with sheep feeding peacefully on the warlike camp, and hawks wivering in the pure air—while North, East, South, and West we gaze over hill, and vale, and down, and woodland that stretch and fade into far distance and vacant haze. So we praise these famous places, fortified by the toil and purpose of our forefathers in the ancient days when Time past unrecorded, except by such earthworks as we have been surveying, and by their makers' castaway possessions.

CASTLE DITCHES (NEAR TISBURY)

CASTLE DITCHES crown the summit of a steep greensand hill that projects from the base of Buxbury Hill. The site commands a long stretch of the Nadder Valley, and is about a mile and a half from Tisbury. Approached from this side, the camp defences are concealed by trees and underwood. A steep wooded hill, fringed along its flat top by wind-blown silver firs and larches, is all that can be seen. The woodland track that leads up to the area is abrupt, deeply sunken, and strongly defended. This approach seems to have been the water-way of the camp dwellers, for the springs rise at the bottom of the hill, where the greensand rests upon gault. On three sides the natural scarps of the hill have needed comparatively little work so as to shape them into a triple semicircle of defensive ramparts. But on the South-Eastern side the approach is flat, and across this plateau of easy access the entrenchments are multiplied. They are most perfect at the extreme Eastern side. Here, there are three great banks and ditches, reinforced by three outer banks and ditches of smaller size, protecting a winding entrance to the area—the whole of which is under cultivation.

On the South-West side of this entrance there are now only four banks and ditches, cultivation apparently having destroyed the two outermost defences. Sir R. C. Hoare gives a plan of Castle Ditches in " Ancient Wiltshire," but it does not show this multiplication of entrenchments on the South-Eastern side. Presumably, as at Whitsbury, he was misled by the woodland growth—for these banks and ditches are concealed by underwood and by timber trees. When I made the section shown on the plan, the underwood had just been cut, which gave a fortunate opportunity for surveying this part of the camp.

All marks of habitation—if any existed—either within the area, or without, have been effaced by cultivation. A long mile of level upland connects Castle Ditches with Buxbury Hill—which is a projecting spur of the White Sheet Hill ridge, and the bank and ditch that cross Buxbury Hollow suggest defence against an enemy coming from these downs towards the camp.

Castle Ditches enclose one of the finest camps within the area of my survey. I think we may regard it as a British tribal centre, and pre-Roman. It is unfortunate in being concealed by woodlands, and in having such an indistinctive name. The place-name of Spelsbury (now forgotten), marked near the camp site in " Andrews and Dury's Map of Wiltshire," 1773, and in the similar map contained in Gough's edition of " Camden's Britannia," 1789, suggests that this earthwork may formerly have been thus named. It may be noted that in neither of the maps mentioned above is this great camp marked, while both record the lesser camp of Chiselbury. A wood is shown in " Andrews and Dury's " map on the hill-top within the area of Castle Ditches, so probably the site was even more concealed then than now.

BADBURY RINGS

Of the five principal camps within the district of my survey, namely, Hod, Hambledon, Castle Ditches (near Tisbury), Whitsbury Castle Ditches, and Badbury Rings, the last stands lowest above the sea; yet Badbury Rings are so isolated, and are situated in such a spacious tract of surrounding lowland, that their pine-crowned summit of 327 feet tells as a landmark for miles around—a distinction missed by Castle Ditches (near Tisbury), though the area of this camp rises to 630 feet—7 feet higher than Hambledon Hill.

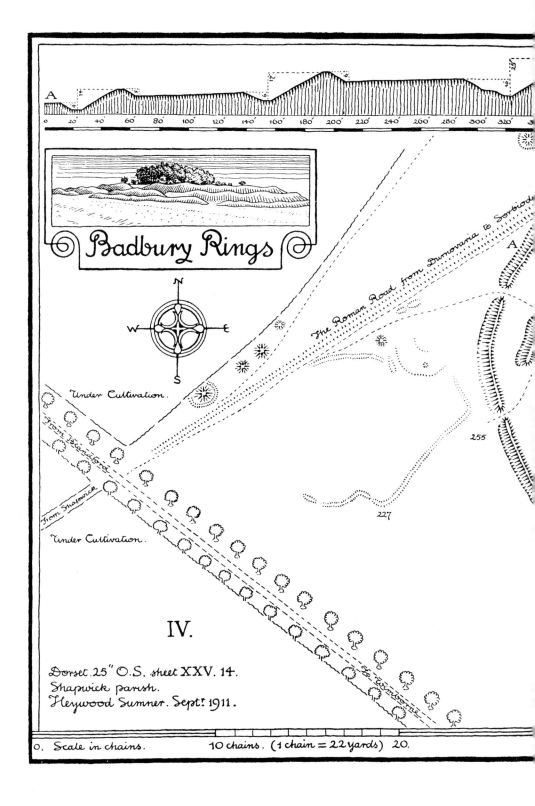

Badbury Rings

IV.

Dorset. 25" O.S. sheet XXV. 14.
Shapwick parish.
Heywood Sumner. Sept! 1911.

0. Scale in chains.   10 chains. (1 chain = 22 yards)   20.

The Roman Road from Vindocladia to Sorbiodunum

Under Cultivation.

From Blandford

From Shapwick

Under Cultivation.

To Wimborne

255

227

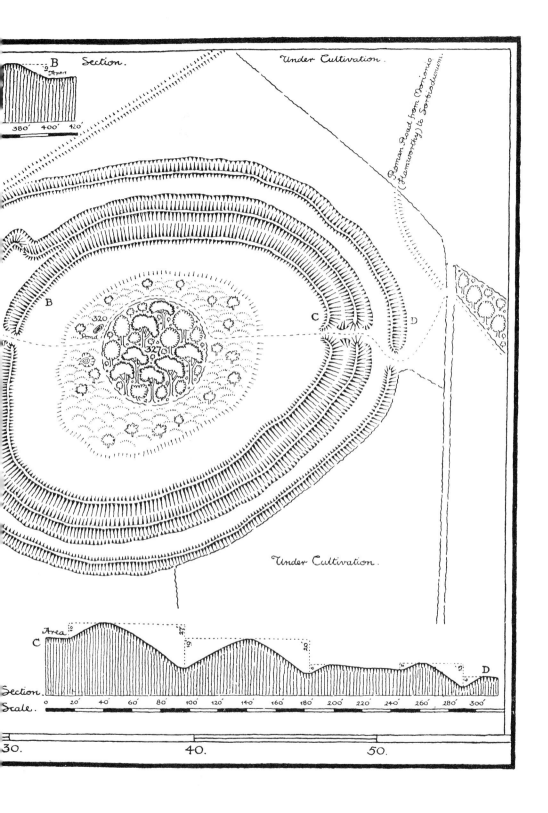

Section.

B

380' 400' 420'

Under Cultivation.

Roman Road from (Vortorio
(Hamworthy) to Sorbiodunum.

B

320

327

C

D

Under Cultivation.

Area

C   D

Section
Scale.

0   20'   40'   60'   80'   100' 120' 140' 160' 180' 200' 220' 240' 260' 280' 300'

30.   40.   50.

BADBURY RINGS

Badbury Rings have been described in the " Dorset Field Club Proceedings," vols. xi and xxvii, and in "Ancient Dorset," by Charles Warne. So far as I know, their varied occupation has not been proved by excavation, but their origin has generally been accepted as pre-Roman. They are surrounded by a triple ring of great banks and deep ditches. There is a wide space between the outermost ring and the two inner rings. The Eastern entrance is a straightforward passage through the three lines of entrenchment. There are two entrances on the Western side. One, winding through the middle ring, that has a berm-like projection at this place; the other, leading straightforward through the Rings. This latter has been doubted as an original entrance, but it seems possible (in view of Mr. and Mrs. Cunnington's excavations on Knap Hill, near Devizes [1]) that it may be original, and may have been used for driving cattle into the area at times of danger, and then these entrance gaps in the rings would have been stockaded.

This camp shows no signs of Roman adaptation, though the site was probably used, or occupied, by the Romans, for three of their roads converge here, namely, the main road from Sorbiodunum (Sarum) to Durnovaria (Dorchester), called Ackling Dyke, as it crosses the downs towards Woodyates; a road to Morionio (Hamworthy); and a branch road traceable across Abbeycroft Down, Hogstock, Launceston, Eastbury Park, Bussey Stool Park (see Plan XIII), Ashmore Common, and lost after Donhead Hollow, where it is last traceable, pointing towards Grovely Ridge.[2] The wasted earthworks outside the Rings, on the Western side, do not seem to have any connection with the camp defences.

Dr. Guest, in his essay on the " Early English Settlements in South Britain " ("Origines Celticae," vol. ii), gives weighty reasons for his conjecture that Badbury was the site of the siege and battle of Mons Badonicus, A.D. 520, recorded by Gildas—a battle which stayed the onward western advance of the West Saxons for nearly half a century. Dr. Guest's reasoning is convincing to one who knows this country and the great fortresses and forests that lay between Charford on the Salisbury Avon, and Bath on the Bristol Avon (which latter site has been claimed as Mons Badonicus). It is difficult to imagine that within a year of the battle of Charford, A.D. 519, the West Saxons could have advanced sixty miles inland, across the strongly fortified downs of Cranborne Chase, through the dense

[1] " Knap Hill Camp," by Mrs. M. E. Cunnington (" Wilts Archaeological Magazine," vol. xxxvii).

[2] See "Some Vestiges of Roman Occupation in Dorset," Rev. J. H. Austen—"Journal of the Archaeological Institute," vol. iv, 1867; and " Roman Road from Badbury to the Wiltshire Boundary near Ashmore," J. C. Mansel-Pleydell—"Dorset Field Club Proceedings," vol. ix, 1888.

forests of Blackmore Vale, past the Great Camp at Cadbury, and so on to Bath. A defeat, then, at Bath, must have meant annihilation to the invaders—whereas a defeat at Badbury would have merely meant a retreat to the Avon, twelve miles distant to the East, and to the New Forest borders, which presumably had at that time been conquered by the West Saxons, and where they were within easy access of the sea—and this is what actually appears to have happened.

WHITSBURY CASTLE DITCHES

THIS is a very fine camp. It stands 400 feet above the sea, its area covers about 16 acres, and it is for the most part surrounded by a triple circle of great banks with two deep ditches. It is the strongest earthwork that commands the lower part of the Avon Valley. But although Whitsbury Castle Ditches are deep and although its banks are high, they are concealed; the road as it winds up the steep hill from Whitsbury Village being sunken, for the first part of the way, in a cutting, and for the remainder, in the outer ditch of the camp; while up and down, the ramparts of the camp are overgrown with timber trees and underwood. Their strength is not readily apparent from any point of view. The woodland growth that covers this earthwork accounts for its scant estimation as a local landmark, and it must have been owing to this concealment that Sir R. C. Hoare described " Whichbury Camp " as "Single-ditched," and that it was omitted from " Andrews and Dury's Map of Wilts," 1773, and from the map of Wilts in Gough's edition of " Camden's Britannia," 1789.

The area, which is now under pasture, presents a smooth surface. There is no humpy ground, but potsherds may be found, telling of past habitation. Probably the area was under cultivation before it was laid down to grass, and the soil of the inner Northern bank that has been partially destroyed was used to chalk the land—a practice that also accounts for the destruction of the Western end of Giant's Grave on Breamore Down close by. It may be noted that the Reading Beds crown Whitsbury Hill with a very shallow cap, according to the proof given by the upturned banks, which are of a chalky nature.

The inner encircling bank is much higher on the South-Eastern side, as though it was intended specially to guard against an attack coming from the plateau ridge towards Whitsbury Common, which slopes upward from the Avon Valley. This inner bank differs in construction from the two outer banks. It rises to a narrow top, and it is much higher—3 feet

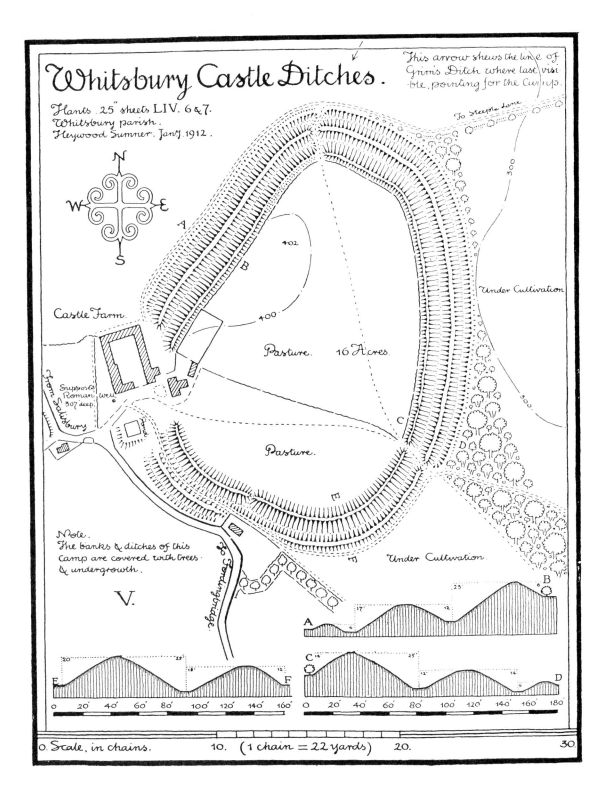

Whitsbury Castle Ditches.

Hants. 25″ sheets LIV. 6 & 7.
Whitsbury parish.
Heywood Sumner. Jan.y 1912.

This arrow shews the line of Grim's Ditch where last visible, pointing for the Currups.

To steeple lane

Under Cultivation

402

400

Castle Farm.

Pasture.   16 Acres.

Supposed Roman well. 30ʹ deep.

From Salisbury

Pasture.

Under Cultivation.

Note.
The banks & ditches of this camp are covered with trees & undergrowth.

To Fordingbridge

V.

A   17   23   B
E   20   25   12   F
C   25   12   14   D

0   20ʹ 40ʹ 60ʹ 80ʹ 100ʹ 120ʹ 140ʹ 160ʹ 180ʹ

0. Scale. in chains.   10.   (1 chain = 22 yards)   20.   30.

across on the top and 25 feet high from the bottom of the ditch, while the outer banks are from 7 to 9 feet across on the top, and 13 to 16 feet high from the bottom of their ditches. Possibly the inner bank may have been added or strengthened in post-Roman times. Its scale is similar to the highest part of Bokerly Dyke, which was proved by General Pitt-Rivers' excavations to have been thrown up after the Roman evacuation.

The long check—from A.D. 520 to A.D. 552—to the Saxon advance westward after the battle of Mons Badonicus (Badbury?), may have been the period when, as I conjecture, this camp was strengthened. Only excavation could prove this supposition, but the shire boundaries here, the Saxon Chronicle, the Moot at Downton, and the place-name of Britford, all suggest that for a time the Avon valley was the boundary between the West Saxon invaders and the Romano-British. Under such circumstances we might expect to find additions made to the defences of such an outlying hill-top camp as Whitsbury Castle Ditches.

The entrances on the North-Eastern and South-Eastern sides of the camp may be original. Probably the main entrance was through the Western gap where now stands the house of Castle Farm.

Grim's Ditch can be traced at the North-Eastern corner of the camp, coming across the arable land from the Shoulder of Mutton Clump at the bottom of Breamore Down. At the Eastern corner, in the hedgerow that grows beside the field track leading down to Steeple Lane, another similar ditch can be traced pointing towards Mizmaze Hill. About 200 yards from the bottom of the hill, it is very clearly defined in the hedgerow beside the field track. Three sentinel yews mark the place to which I am referring. There are no signs of the bank or ditch across the arable land to the East of Steeple Lane, but on Breamore Down they can be traced winding up the hill, past the Giant's Grave, to the pond, and then round the Mizmaze Hill until they are lost on the North-Eastern side. A third ditch is shown on the Ordnance Survey map proceeding from the South-Eastern side of the camp towards Whitsbury Common, but its trace has now been destroyed by cultivation, except within an oak and hazel copse called Rowdidge (or Rowditch).[1] Here a wasted bank and ditch can be traced through the copse wood (a few yards to the left of the woodland track), and continue over the ridge, dying away in Newland Bottom—towards Roundhill Farm.

On the Western side of the camp formerly stretched one of the old Parks of Hampshire. Whitsbury Park is shown in the maps of Saxton, 1579, Norden, 1595, and Speed, 1610. Probably both waste and wood-

---

[1] The name Row-ditch recurs on the White Sheet Hill Ridge, near the British settlement on Swallowcliffe Down.

land had to be grubbed up before this land could be brought under its present state of cultivation, and there are no signs to be now seen of bank or ditch crossing from Whitsbury towards Rockbourne and Damerham Knoll. The convergence of these boundary banks and ditches towards certain centres of safety and habitation is a point to be noted in the survey of pre-Roman camps and settlements. Similar convergence of such earthworks may be seen at the British settlements on Gussage Down, on Blandford Race-course, near Bokerly Dyke (South-West side), on South Tarrant Hinton Down, Middlechase Farm, and at Buzbury Rings. It seems possible that these boundary banks and ditches were needed to turn the cattle towards the camps of safety when they were hurriedly driven, in times of emergency. Or that they were needed for purely pastoral usage in peaceful periods, when the camps may have been used as cattle pounds, instead of as defences. In support of this latter supposition we have to note the open undefended sites of the Romano-British settlements on Cranborne Chase, which certainly suggest Pax Romana, and a period when there was no need for defensive entrenchments.

WINKELBURY

THIS camp was partially excavated by General Pitt-Rivers in 1881-2, and the results are recorded by him in "Excavations in Cranborne Chase," vol. ii. The evidence obtained from his investigations led him to think that it was probably a pre-Roman camp of habitation, specially constructed to allow for the hasty collection of cattle in the event of a raid; and that it was very little, if at all, occupied by the Romans. It was a hardy, poor place of pastoral habitation, and it bears no traces of having ever been occupied by a race of men who had reached any advanced stage of civilization.

The Northern, or lower division of the camp appears to have been used for habitation, and the Southern or upper part of the area for cattle. The wide gaps in the defences on the Southern side were proved by excavation to be original, and they were presumably required for the hasty inlet of cattle when they were herded to escape from raiders.

The sunken tracks on the Eastern and Western sides of the hill suggest traffic to and from the sources of water supply, for springs rise in the valleys on either side.

The plan will show the curious projecting character of the hill on which this camp is situated. On three sides its scarps are abrupt, while

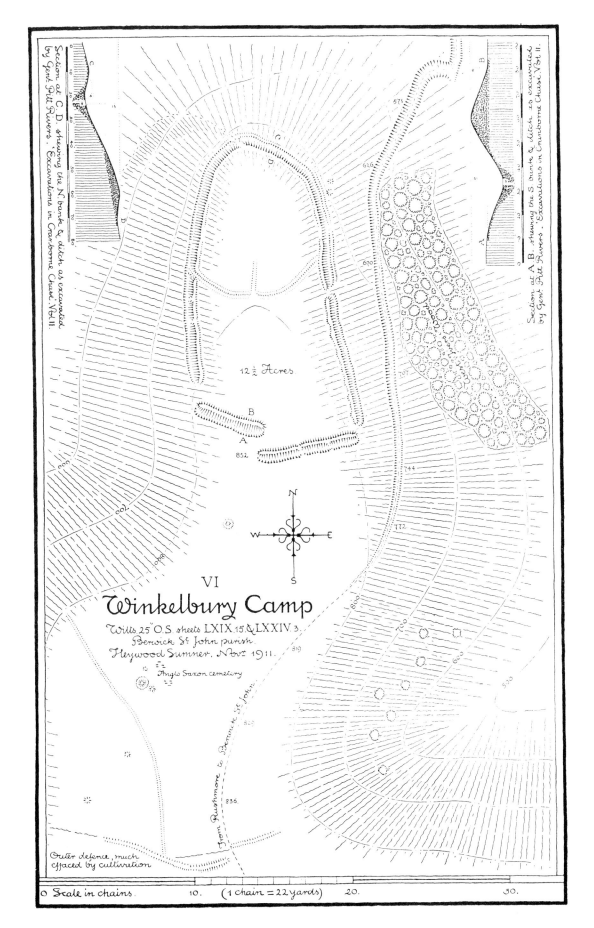

Section at C.D. shewing the N. bank & ditch as excavated by Genl. Pitt Rivers. Excavations in Cranborne Chase. Vol II.

Section at A.B. shewing the S. bank & ditch is excavated by Genl. Pitt Rivers. Excavations in Cranborne Chase. Vol II.

12½ Acres

B
A
852

744

772

800

819

636

500

600

700

N
W E
S

VI
Winkelbury Camp
Wilts 25" O.S. sheets LXIX.15.& LXXIV.3.
Berwick St. John parish.
Heywood Sumner. Nov. 1911.

Anglo Saxon cemetery

from Rushmore to Berwick St. John.

Outer defence, much effaced by cultivation

O Scale in chains.   10.   (1 chain = 22 yards)   20.   30.

on the Southern, upland side, where it joins the Oxdrove Ridgeway, there are the remains of a scarp to scarp entrenchment that must have further added to the security of this pastoral hill corner with a wide opening near the Eastern scarp. Winkelbury—Wincel-burh, A.S. *The Corner Fort*—a name that truly describes its remarkable position.

Long days spent in wandering over these uplands of Cranborne Chase fix the vision of their summits of vantage in the mind's eye—and sometimes we may question the choice of our prehistoric forefathers. Certainly at Winkelbury they chose a defensible hill spur, but, rather unusually, they also chose a site sloping to the North for their pastoral camp of habitation. It seems strange that Win Green, two miles to the West, should bear slight traces of primitive occupation, for it is the highest point of these downs (911 feet above the sea), it commands the grazing of Charlton Down on the Western side, it has a long spur trending South towards Tollard Royal, and it is connected with the Upper Bridmore leaze by a narrow defensible ridge. Probably water supply was the reason that led the British herdsmen to choose Winkelbury rather than Win Green, but we miss the earthen record of "the great days done" as we stand on this bare supreme height of Cranborne Chase and gaze across miles and miles of lesser heights that were crowned by the Briton's choice.

This is a digression, however, from Winkelbury on which we ought to be standing. Though Win Green is pre-eminent, Winkelbury Camp is a fine site. It dominates the springs of the Ebble Valley, and it overlooks the Vale of Wardour, with Shaftesbury, Tittlepath Hill, and White Sheet Hill in the middle distance, while beyond lie Grovely Ridge and long lines of downland that mark the expanse of Salisbury Plain.

CASTLE RINGS

THE camp known as Castle Rings is situated near the Eastern end of the upper greensand ridge whereon Shaftesbury stands. The entrenchments are entirely hidden by undergrowth, and the passer-by will probably not turn aside to see the camp that is surrounded by hedgerow thickets of hazel, elder, and oak. These cover the banks, while brambles and wild raspberries help to conceal the ditch. Only when the leaves are off can any idea be formed of the defences that surround Castle Rings.

A single ditch between double banks encloses the camp, and the banks have been partly cut away by cultivation. Throughout, the earthwork is continuous and regular in construction, though at the Northern side the ditch is now rather deeper and wider than elsewhere. The area is under cultivation. In Sir R. C. Hoare's plan of one hundred years ago, it is shown as divided with hedges into three fields; now, it is thrown into one. All signs of habitation or of user have been effaced by cultivation, and we have no clear indications as to the probable period of this camp. There are four entrances, North, South, East, and West —merely gaps in the earthworks which may or may not be original.

In the triangular field on the South-West side Sir R. C. Hoare, in " Ancient Wiltshire, South," shows an outlying bank and ditch defence to the camp. This is now spread by cultivation, but it can still be traced across the arable field between the two roads, and across the grass strip on the North side of the Semley road; then in Crates Wood on the North side of the hill it is clearly defined, though partially hidden by young firs and undergrowth. On the South side of the Donhead road this outer bank and ditch can also be traced, as it dies away on the steep wooded scarp of the hill-side. The ditch is on the Western side, and the defence appears to have been thrown up against an enemy coming from the direction of Shaftesbury and of Blackmore Vale, i.e., from the West. The usual line of attack elsewhere on Cranborne Chase seems to come from the East. This exception may be explained by the natural position of Castle Rings, which are situated on an isolated returning ridge of high ground that runs parallel with Win Green and Charlton Down. An enemy coming from the East along the Oxdrove Ridgeway would probably have been obliged to follow the high ground over Win Green and Charlton Down to Shaftesbury, and would then have to turn back towards the North-East to attack Castle Rings. This appears to be a possible explanation of the arrangement of this outlying bank and ditch that we are considering.

The view is most beautiful over Blackmore Vale towards Mere, Stourhead, and Alfred's Tower, on the North-West side, and over Wardour Vale towards Whitesheet Hill and Win Green on the Eastern side. In the foreground, deep fertile valleys wind round little wooded hills; the rich pastures of Blackmore Vale creep up to the ridges that encircle the lowlands, and long lines of downland stretch their bare curves against the far horizon.

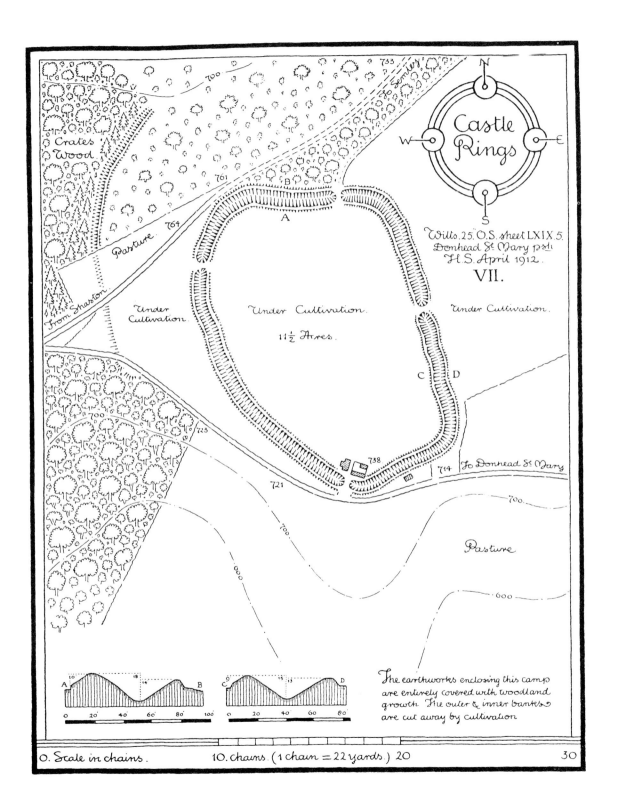

Castle Rings

N
W   E
S

Wilts. 25. O.S. sheet LXIX 5.
Donhead St Mary psh
H.S. April 1912.
VII.

733
Semley
700
761
Crates
Wood
764
Pasture
From Shaston

B
A
Under
Cultivation

Under Cultivation.

11½ Acres.

Under Cultivation.

C D

700
723
738
721
714
To Donhead St Mary
700

600

600

Pasture

A 10 18 B
0 20 40 60 80 100

C 9 15 13 D
0 20 40 60 80

The earthworks enclosing this camp
are entirely covered with woodland
growth. The outer & inner banks
are cut away by cultivation

0. Scale in chains.   10. chains. (1 chain = 22 yards.) 20

30

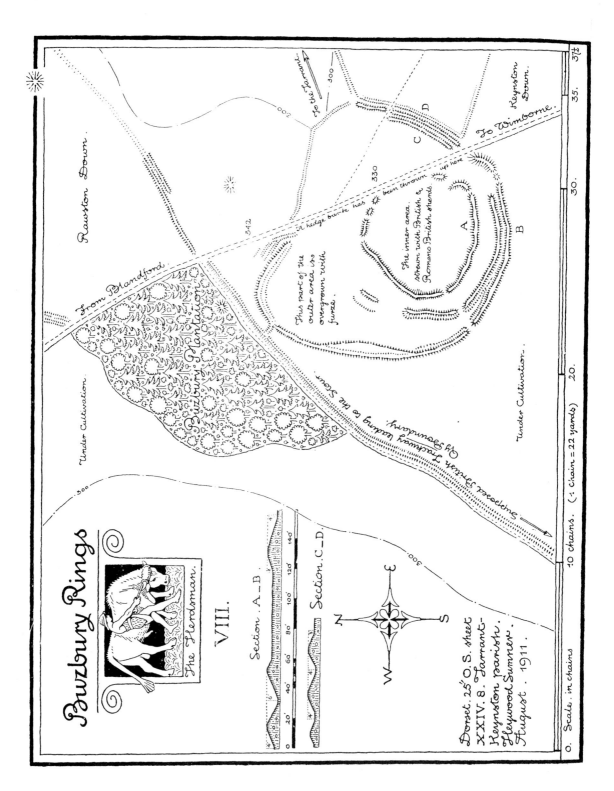

Buzbury Rings

The Herdsman.

VIII.

Section . A – B .

Section . C – D .

Dorset. 25" O. S. sheet
XXIV. 8. Tarrant-
Keynston parish.
Heywood Sumner.
August . 1911.

Scale . in chains

0. 10 chains. (1 chain = 22 yards) 20. 30. 35. 37½.

Rawston Down .

From Blandford .

Buzbury Plantation .

Under Cultivation .

542

300

200

330

300

This part of the
outer area is
overgrown with
furze.

The inner area .
strewn with British &
Romano British sherds

a hedge bank has been thrown up here

Old trackway, leading to the Stour

Rawston Down .

To the Tarrant

To Wimborne .

Keynston
Down .

Under Cultivation .

A

B

C

D

BUZBURY RINGS

BUZBURY RINGS are about two miles distant from Blandford, and the upland road, thence to Wimborne, passes through the outer part of the camp. The inner camp appears to have been the place of habitation, and here potsherds of various kinds may be picked up on the rabbit-scrapes and mole-hills. Many of these sherds seem to be of the early British type, handmade, imperfectly baked, and made of clay mixed with siliceous granules—suggesting an early occupation. The outer camp extends on the Northern and Eastern sides, and shows no signs of habitation, but was probably used for pastoral purposes. Buzbury Rings have been cut about by road-makers and by cultivators, but their general disposition is still fairly discernible. The camp shows no signs of having been strengthened, and its broad-topped low banks (six feet average height) and shallow ditches give us an idea of a British tribal camp that combined safety with pastoral requirements.

The site is pastoral rather than defensive, and if desperate emergency arose, it is intelligible that Badbury Rings (five miles to the South) or Hod Hill and Hambledon Hill (five miles to the North) would be chosen as camps of safety, or as camps to be strengthened rather than Buzbury. I do not think that this earthwork played much part in resisting the West Saxon advance westward in the sixth century.

Banks and ditches of the Grim's Ditch type converge on Buzbury Rings. The Ordnance Survey marks one that approaches from Langton Long as "Supposed British Trackway." It continues on the North-Eastern side of the Blandford and Wimborne Road, pointing towards Blandford Race-down. Another appears to lead towards the Tarrant Valley, and another may be traced, under cultivation, towards Spettisbury Ring. The all-over surface measurements of these continuous banks and ditches compare with those through which I have cut sections on Knoll Down and on Gallows Hill. They would appear to be boundary divisions for pastoral purposes. Similar convergence of such earthworks towards centres of British habitation may be noted at Whitsbury Castle Ditches, and at the British settlements on Blandford Race-down, South Tarrant Hinton Down, Gussage Down, near Bokerly Dyke (South-West side), and Middle Chase Farm.

CHISELBURY CAMP

CHISELBURY CAMP stands on the broad plateau ridge that separates the Nadder Valley from the Ebble Valley. Its circular area, now under cultivation, is enclosed by a single bank and ditch that are very regular in construction. The entrance is on the South-Eastern side, and it is covered by a bastion outer defence that has been cut about and partially effaced by cultivation, and which compares with a somewhat similar out-work at the entrance of the enclosure above Chickengrove Bottom. On the North side of the camp, a ditch between two banks (of the Grim's Ditch type) issues from the entrenchment and crosses the down to its steep scarp edge, where it dies away. On the South side a ditch issues from the outer bastion defence, and runs down to the Salisbury and Shaftesbury Road as a ditch only. On the South side of this road it reappears, but now as ditch between two banks, similar to the earthwork on the North side, and continues thus for the few intervening yards between the road and the head of the abrupt combe, where it dies away.

There is also a row of three banks and two ditches farther to the West, and South of the road; but the down has been much cut about here, and I doubt the antiquity of this triple row. It is unfortunate that the area of the camp is under cultivation, as it prevents us from forming any opinion respecting the possible continuity of these North and South ditches. They may be the remains of a scarp to scarp boundary earth-work that crossed the down before the camp was made.

When Sir R. C. Hoare made his survey, the greater part of the camp was covered with heath and furze. He dug in several parts of the area but discovered no bones, pottery, or signs of ancient habitation ("Ancient Wiltshire—South," p. 249). From the form and regularity of the work, he was inclined to think this camp to have been one of comparatively late date.

This isolated chalk ridge, on which stands Chiselbury Camp, runs East and West from Harnham Hill above Salisbury, to White Sheet Hill above Berwick St. John. It is thus described by Dr. Stukeley in his "Itinerarium Curiosum," 1724, 2nd edition 1776, p. 135: "The road " from Wilton to Shaftesbury, called the Ten-mile Courſe, is a fine ridge " of downs, continued upon the ſouthern bank of the river Nader, with a " ſweet proſpect to right and left, all the way, over the towns and the " country on both ſides: a traveller is highly indebted to your lordſhip " (Lord Pembroke) for adding to his pleaſure and advantage, in reviving " the Roman method of placing a numbered ſtone at every mile, and the " living index of a tree to make it more obſervable; which ought to be " recommended as a laudable pattern to others. . . . Between No. 5 and " 6 is a pretty large camp, called Chiſelbury, upon the northern brow of " the hill: it is ſingle ditched and of a roundiſh form: before the chief " entrance is an half-moon, with two apertures for greater ſecurity: there

26

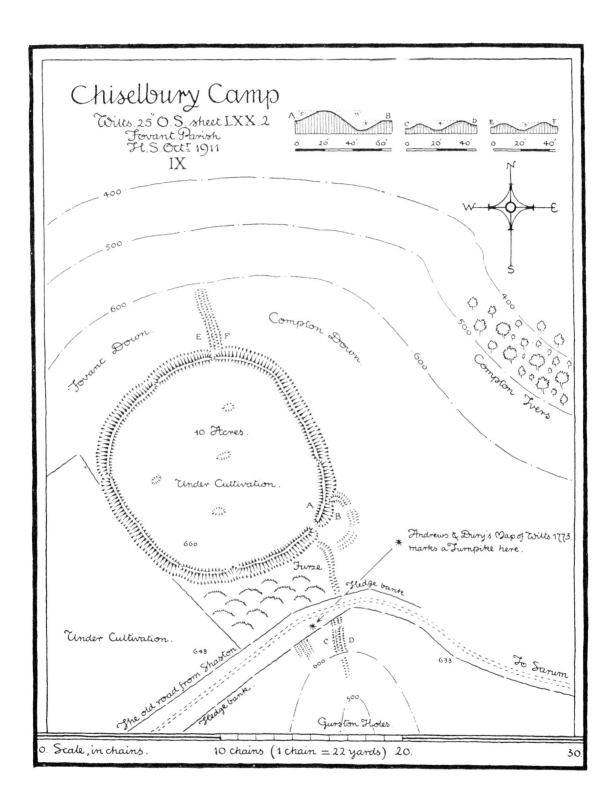

Chiselbury Camp

Wilts. 25" O.S. sheet LXX.2
Fovant Parish
H.S. Oct. 1911

IX

400

500

600

Fovant Down

Compton Down

Compton Ivers

E F

10 Acres.

Under Cultivation.

660

A

B

Furze

Andrews & Dury's Map of Wilts 1773.
marks a Turnpike here.

Hedge bank

Under Cultivation.

648

The old road from Shaston.

Hedge bank.

C D

600

633

To Sarum

500

Gurston Holes.

0. Scale, in chains.   10 chains (1 chain = 22 yards) 20.   30.

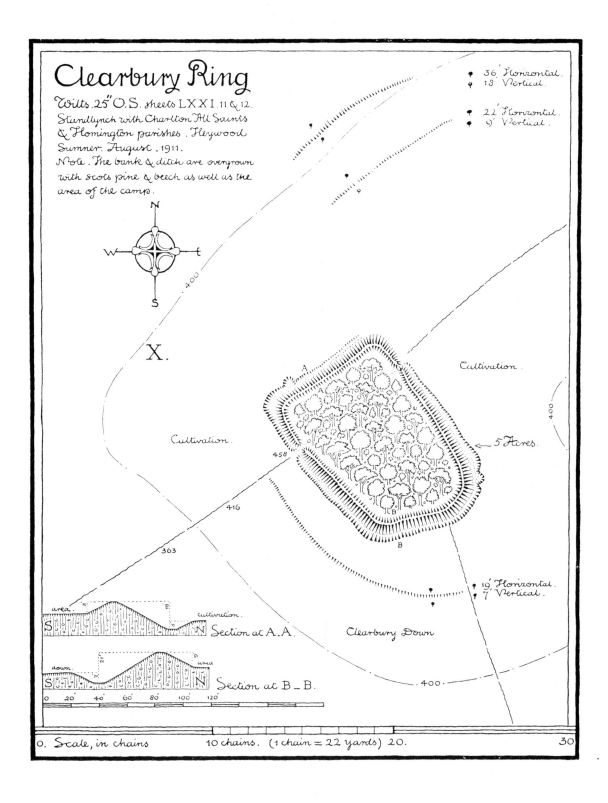

Clearbury Ring

Wilts. 25" O.S. sheets LXXI.11 & 12.
Standlynch with Charlton All Saints
& Homington parishes. Heywood
Sumner. August. 1911.
Note. The bank & ditch are overgrown
with scots pine & beech as well as the
area of the camp.

36' Horizontal.
18' Vertical.

22' Horizontal.
9' Vertical.

X.

Cultivation.

Cultivation.

5 Acres.

458

416

363

19' Horizontal.
7' Vertical.

area.   cultivation.
S   N Section at A.A.

down   area
S   N Section at B _ B.

0 20' 40' 60' 80' 100' 120'

Clearbury Down

400

" is a ditch indeed goes from it downward to the valley on both fides,
" but not to be regarded. This I imagine relates not to the camp; for I
" obferved the like acrofs the fame road in many places between little
" declivities, and feem to be boundaries and fheep-walks made fince, and
" belonging to particular parifhes." While doubting Dr. Stukeley's
imaginings as to these bank and ditch barriers, we may value his observation. The "Ten-mile Course" is now a broad drove way along which the
traveller may fare from its rise on Harnham Hill to its fall on White Sheet
Hill, with only sheep, plovers, and hares to break the solitude. Once
upon a time this was the highway from Salisbury to Shaftesbury, and a
turnpike stood close to Chiselbury Camp ("Andrews and Dury's Map of
Wiltshire," 1773). The ridge bounds the Nadder Valley to the North
with abrupt and continuous scarps of downland, and to the South with
upland spurs, projecting at right angles into the Ebble Valley, and intersected by deep combes—the uplands being under cultivation, and the
combes downland. As this ridge runs onward to the West, it rises from
300 feet above the sea at Salisbury racecourse, to 700 feet at White Sheet
Hill, while the Northern scarp becomes continuously steeper, and the
ridge spurs and combes on the Southern side become more sharply
defined. Chiselbury is the only defensive camp, and Swallowcliffe Down
the only British settlement on this long upland; but these earthworks
must not be regarded as the measure of its occupation, for there has been
much cultivation on these hill-tops, with the result that we only obtain
a hint of its prehistoric occupation.

The old name of "The Ten-mile Course" seems now to be quite
forgotten.

CLEARBURY RING

On a high hill south of *Hummington* or *Odstock* is a very great single camp
called *Clerebury* with a beacon in it (Gough's edition of "Camden's Britannia,"
1789).

Clearbury Ring is a mean earthwork when compared with the very fine speci-
mens which our county has afforded, but it stands pre-eminent in point of extensive
prospect, and is seen at a very considerable distance (Sir R. C. Hoare, "Ancient
Wiltshire—South," 1812).

OF these two descriptions, Sir R. C. Hoare's is the better. It is owing
to its eminent and isolated position rather than to its being "very great,"
that Clearbury Ring is recorded in the earliest maps which noticed earthworks, namely, "Andrews and Dury's Map of Wiltshire," 1773, and that

contained in Gough's edition of "Camden's Britannia," 1789—both of which, however, omit the really "very great" camp at Whitsbury close by.

The "Ring" is defended by a single big bank and ditch, and there are gap entrances on the North-East and South-West, of which the latter appears to be original. The area of the camp only covers five acres, and is oblong in shape. Apparently the early cartographers above-mentioned assumed its shape from its name, for they mark Clearbury Ring as a precise circle. It is surrounded by the remains of cultivation banks, but there are no signs of habitation on the adjacent down.

The clump of beech and Scots pine that crown the hill adds greatly to its beauty and significance in the landscape, but the archaeologist must regret that the woodland growth, over the camp site, prevents the hope that its history may be unearthed by the spade. The periods of the occupation of Clearbury Ring; the connection between it and the fragmentary camp in Odstock Copse; and the possibility of the bank defence having been added to at some time of emergency (query, the West Saxon advance from Downton)—these are questions that excavation might here have answered without excessive search and expenditure of labour, the area of the camp being comparatively small. But now the roots of trees hold the secrets of Clearbury Ring.

DAMERHAM KNOLL

DAMERHAM KNOLL—like Pentridge—is an outlying chalk hill, capped with Reading Beds, and accordingly the wasting nature of its top soil has brought about the partial effacement of its earthworks. The knoll is only 429 feet high, but owing to the lie of the surrounding land and to its isolation, it is a prominent landmark for miles around. The camp is small—the area includes about three acres—is on the summit of the knoll, and is surrounded by a ditch with double banks. In their present condition these defences would scarcely be noticed by a passer-by. They are strengthened on the North-Eastern side by an outlying bank and ditch which follow the scarp of the hill, and which, after crossing at right angles the boundary ditch that runs from Blagdon Hill to Knoll Hill, die away on the down about seventy yards beyond this intersection, pointing towards Grans Barrow and Knap Barrow; and also by a scarp to scarp bank and ditch which cross the Knoll ridge about fifty yards to the North of the camp entrenchment. It is difficult from superficial

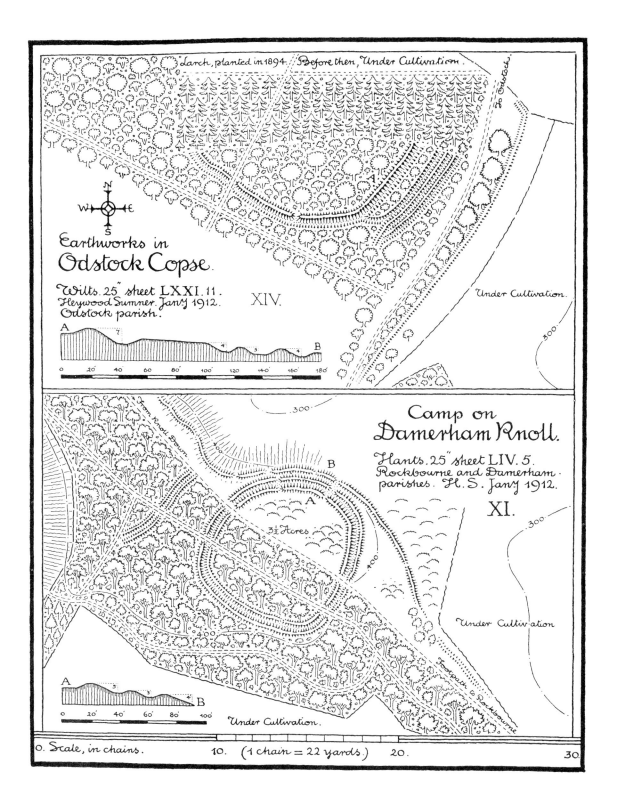

Larch, planted in 1894. Before then, Under Cultivation.

To Odstock

Under Cultivation.

300

Earthworks in
Odstock Copse.

Wilts. 25" sheet LXXI. 11.
Heywood Sumner. Jan'y 1912.   XIV.
Odstock parish.

A — 7 — 4 — 3 — 4 — B
0   20' 40' 60' 80' 100' 120' 140' 160' 180'

300

Camp on
Damerham Knoll.

Hants. 25" sheet LIV. 5.
Rockbourne and Damerham.
parishes. H.S. Jan'y 1912.

XI.

300

3½ Acres

Footpath to Rockbourne

Under Cultivation

A — 5 — 3 — 4 — B
0   20' 40' 60' 80' 100'

Under Cultivation.

0. Scale, in chains.   10.   (1 chain = 22 yards.)   20.   30.

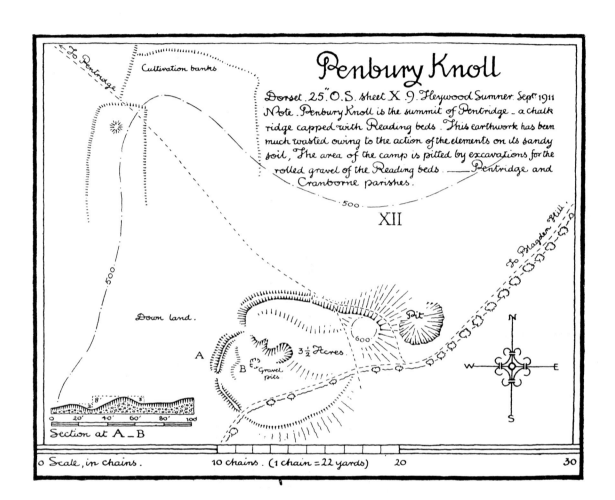

Penbury Knoll

Dorset. 25" O.S. sheet X. 9. Heywood Sumner. Sept.ᵗʰ 1911
Note. Penbury Knoll is the summit of Pentridge – a chalk
ridge capped with Reading beds. This earthwork has been
much wasted owing to the action of the elements on its sandy
soil. The area of the camp is pitted by excavations for the
rolled gravel of the Reading beds. _____ Pentridge and
Cranborne parishes.

To Pentridge

Cultivation banks

500

XII

To Blagden Hill.

Down land.

Pit

600

A

B   Gravel
pits

3½ Acres

N

W   E

S

Section at A _ B

0 20' 40' 60' 80' 100

0 Scale, in chains.   10 chains. (1 chain = 22 yards)   20   30

inspection to be sure of the entrance to this camp. Tree planting and cartways have confused the original design; but it seems to have been outside the plantation of Scots pines, on the line of the present pathway.

In size, this camp on Damerham Knoll compares with those on Penbury Knoll, in Mistlebury Wood, and in Bussey Stool Park. The British potsherds that may be found on the rabbit scrapes and mole-hills of Knoll Down, and the cultivation banks that may be seen on the hill-side towards Toyd Farm, near Grans Barrow and Knap Barrow, testify to ancient habitation and cultivation hereabouts.

This earthwork may have been the camp of refuge in time of emergency for the prehistoric dwellers on Knoll Down. I suspect British habitation both on, or near, Knoll Down, and on Rockbourne Down (near Duck's Nest), but have not felt sufficiently sure, so as to mark these sites as British settlements.

In calling this camp Damerham Knoll I have followed the Ordnance Survey; but it should be noted that the hill is known as Rockbourne Knoll by the inhabitants of Rockbourne.

PENBURY KNOLL

THE top of Penbury Knoll is formed of a sandy cap of Reading Beds interspersed with rolled pebbles. Accordingly these earthworks have been wasted both by the elements and by the burrowing rabbits, while further, the small area of the camp has been scored by the digging of "pobble-stone" gravel pits. So much has the camp been thus wasted and cut about that it is not possible to be sure as to the original entrance, or to form an adequate idea of its defences. In situation and size it compares with the camp on Damerham Knoll, three miles distant, to the North-East. It appears to be a pre-Roman hill-fort camp of refuge for the dwellers on Pentridge, and I do not understand why General Pitt-Rivers considered that it was "a Saxon or Norman Burgh" ("Excavations in Bokerly Dyke," vol. iii, p. 57).

There are the remains of many cultivation banks on Pentridge, with frequent signs of ancient habitation, and also on Blackbush Down, its eastern extremity. In "Excavations in Bokerly Dyke and Wans Dyke," vol. iii, p. 240, General Pitt-Rivers illustrates a fine Bronze Age drinking vessel or "beaker" that was dug up near Blackbush Down.

The view from Penbury Knoll is specially fine. It is the summit (600 feet) of the down ridge—Pentridge—that runs for two miles from

Blackbush to Blagdon—an outlying hill set in the midst of the lower slopes of Cranborne Chase. These lower slopes stretch from Great Yews, Westward, to Blandford Race-down, and they are watered by seven chalk streams that flow Southward to the Avon, and to the Stour. They lie in folds that rise and fall from East to West, with the Oxdrove Ridgeway barring the North and North-Western distance as it slowly mounts to the hill-tops of Win Green, Charlton Down, and Ashmore. To the East rises Damerham Knoll, and to the South lie miles of woodland, heathland, and valley leading to the distant sea.

Indeed a most beautiful point of view, but a wind-swept site for habitation. Here every growing thing proclaims tough tenacity as the standard of life. The thorns—the black bushes—are wizened by the searching winds. No lichen grows on the windward side of their dark and twisted branches. The yews turn towards the East, and thus tell of the prevailing storms from the West. Raw, white patches of chalk and flints show where the rabbits burrow under the straggling elder trees, and the pale yellow bent grass, the moss, and the heather, all starved and stunted, tell the same tale—that life is a grim struggle on Pentridge.

Yet this place was once chosen by our forefathers. Here they made their pit dwellings, hunted, herded their cattle, tilled the ground, lived and died, as tough and tenacious as the thorns, and yews, and elders that still occupy the hill. From here, at least, raiders could be seen from afar, while the steep down scarps guarded the approaches to the camp. And here, too, its inhabitants gained better security against four-footed raiders—the wolves of the woodlands and of the swampy valleys. Pentridge might be buffeted by every wind that blows, might be soaked by driven rain, or scorched during spells of drought, but it was free from the dangers of the lowlands, and its isolated eminence afforded a certain measure of security in prehistoric times.

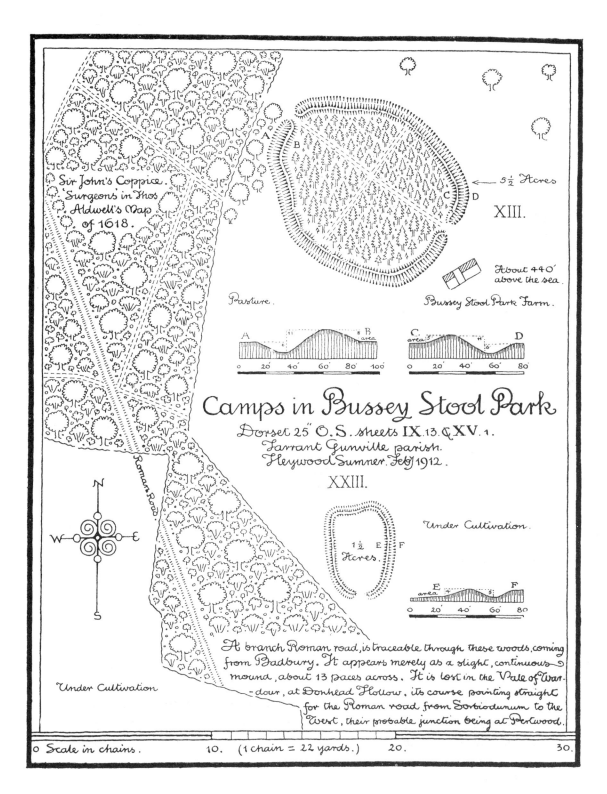

Sir John's Coppice.
Surgeons in Thos
Aldwell's Map
of 1618.

Roman Road

N
W E
S

Under Cultivation

Pasture.

A
B

5½ Acres

XIII.

About 440'
above the sea.

Bussey Stool Park Farm.

A B
area
0 20' 40' 60' 80' 100'

C
area
D
0 20' 40' 60' 80'

Camps in Bussey Stool Park

Dorset 25" O.S. sheets IX.13. & XV. 1.
Farrant Gunville parish.
Heywood Sumner. Feb. 1912.

XXIII.

1½
Acres.
E F

Under Cultivation.

E
area
F
0 20' 40' 60' 80'

Under Cultivation

A branch Roman road, is traceable through these woods, coming
from Badbury. It appears merely as a slight, continuous
mound, about 13 paces across. It is lost in the Vale of War-
-dour, at Donhead Hollow, its course pointing straight
for the Roman road from Sorbiodunum to the
West, their probable junction being at Pertwood.

0 Scale in chains.   10.   (1 chain = 22 yards.)   20.   30.

CAMPS ON HIGH GROUND

THE CAMPS IN BUSSEY STOOL PARK

THESE two camps are on the left of the road from Tarrant Gunville to Larmer Tree—about a mile from the entrance of the Bussey Stool woods at Bloody Shard gate. The larger of the two stands on the top of a level plateau, about 450 feet above the sea. It is enclosed by a broad outer ditch, well defined, from 6 to 8 feet in depth, and by an inner bank that rises 14 feet from the bottom of the ditch on the North-West side, and 8 feet above the level of the area. In many places this encircling bank has been cut about by diggings— possibly owing to ferreting, or to root-grubbing, for the camp was overrun with rabbits, and overgrown with timber before the area was planted with larch as we now see it. Whatever the origin, the result is that the crest of the camp bank is very irregular in many places, but without any suggestion that such irregularity belongs to its original construction, or to its subsequent strengthening. There are two narrow entrances to the camp. One on the North-West side, the other on the South-East side, both of which appear to be original. The entrance on the North-West side is remarkable, for the ramparts turn outwards instead of incurving, as is so frequently the case at the entrances of pre-Roman camps. There is no similar instance on Cranborne Chase, but it compares with the Western entrance through the outermost ring of earthworks at Danebury, near Stockbridge.[1]

This was certainly a defensive camp; indeed its entrenchments are greater than those of any of the lesser camps on Cranborne Chase—excepting Clearbury Ring.

About 250 yards to the West there is a smaller camp that lies on either side of an upland valley that slopes towards the Tarrant. It has quite recently been destroyed, the land having been reclaimed from its previous rough, woodland state, and grubbed and ploughed over in January 1912. The section shown in the plan was made at a place that had been left untouched owing to the plough having been diverted by a tree stool that

[1] See "Danebury," described and planned by Dr. Williams Freeman, "Hants Field Club Proceedings," 1910.

31

remained on the bank. No signs of habitation appear to have been turned up by the plough. I searched the furrows in vain for potsherds, and for cooking flints. The situation, the small size of the bank and ditch, and the wide entrance, all suggest a pastoral enclosure.

This recent example of destruction is suggestive. It shows how easily such destruction may be effected, and in a few years time the recurring processes of cultivation will have effaced all traces of this pastoral enclosure. The site will then suggest the fighting purpose of our forefathers without any hint of their farming pursuits, and it seems probable that such mis-record of primitive life may frequently occur, and for the same reason—the pastoral camps being easily destroyed owing to their low earthworks, while the defensive camps remain, being preserved from the attacks of the peaceful cultivator by the size of their entrenchments.

In Warne's " Ancient Dorset," p. 47, a woodland camp is mentioned which seems to refer to the larger of these two camps. The name, *Hook's Wood*, belongs to a portion of this great woodland near New Town, on the Farnham side, and it appears to have been given to this earthwork which then—in 1872—was more overgrown than at present. This is his description of the camp: " When, however, we come to the works " in Hook's Wood . . . we at once find ourselves in the presence " of an earthwork in every respect analogous to Cæsar's description of " the oppida which he encountered in the woods of Kent and Sussex, and " of which he says ' When the Britons have fortified a thick wood with " ' a bank and ditch, they call it a town, (oppidum), where they are " ' accustomed to assemble when they fear a hostile attack.' . . . In form, " this oppidum is circular, like Badbury and Buzbury, and consists of a " very strong vallum, with a fosse of proportionate depth on the exterior, " affording together a simple but formidable means of protection. It is " pierced by two entrances, one on the N.E. (Qy. S.E.) the other on " the N.W. which are not defended by any supplementary works. The " area is about 180 yards in diameter, and is covered with wood."

Mr. Charles Warne's assumption that this camp was originally set in a forest does not appear to be certain. There is forest land close by—on Bussey's Down—but most of the surrounding woodland is planted—hazel underwood with oaks above—here and there, groves of birch trees, and scattered abeles and spruce firs. The survival of an ancient forest is not certainly shown by the woodland growth. Moreover the precise subdivisions and names given to these woods in Thos. Aldwell's " Mappe of Cranburne Chase," 1618 (" Great Ditch " and " Little Ditch " appear as place-names suggesting these two camps), show that these woods were tended and probably planted in mediaeval times. I do not think that we can be sure that Bussey Stool Park, South Lodge, Rushmore, Mistleberry, or Odstock Copse camps, all of which are now surrounded with woods, were originally woodland camps, or oppida, such as Caesar described.

The pastoral camp mentioned above appears to have been overlooked by Mr. Charles Warne, as he does not refer to it.

ODSTOCK COPSE

The remains of the camp in Odstock Copse (see plan facing p. 28) are of considerable interest. They may be found at the Northern end of the copse, to the right of a woodland ride that runs from the grass Avenue track to the Whitsbury and Odstock road. The double banks and ditches of the outer berm defence are close to the Avenue track, but are concealed by the copse. It is impossible to survey these entrenchments in summer-time, for they are completely hidden from view when the thickets of undergrowth are in leaf. But in winter, or in early spring, some idea may be gained of the remnant of this camp, and of the remarkable berm defence on the Eastern side.

The camp was enclosed by a single bank and ditch, except where the berm projects, and here two more banks and ditches were added to its defence. No original entrance can be traced in the fragment that remains. Its position is much less defensible than that of Clearbury Ring, which stands on an isolated hill a hundred feet higher than Odstock Copse and about a mile distant to the East.

The area of this camp can only be guessed, but it seems probable that it contained about ten acres, that is to say, twice the acreage of Clearbury Ring. The fragment that remains of this earthwork does not appear to have been added to. Towards Little Yews, on the South-West, there are many cultivation banks, and likewise to the East, on the Clearbury Ring side. The site suggests considerable and constant occupation in ancient times, but cultivation, and subsequent tree planting have caused such changes of the surface that a survey of this camp fragment can only result in conjecture. The different construction of this entrenchment from that of Clearbury Ring is noticeable, as their respective sections testify.

It is unfortunate that both these sites—Odstock Copse and Clearbury Ring—should be overgrown with trees, and that we cannot hope that the excavator's spade may unlock their secrets.

It is specially unfortunate because there seems ground for supposing that the West Saxon advance may have been along this line. From 520 to

552 the West Saxon advance appears to have stopped on the East bank of the Avon. In 552 the battle of Searo-byrig (Old Sarum) marks a forward move, and in 556 the battle Berin-byrig (Barbury Hill?) a still further advance, leading up to the battle of Deorham, 577, near Gloucester, when, if not before, the fate of the Cranborne Chase district must have been sealed. And this victorious advance may have begun from Downton.[1] Accordingly the camps at Clearbury and Odstock Copse must be marked with a large note of interrogation, which I fear will stand.

MISTLEBERRY WOOD

This camp is small—about 2 acres—strongly entrenched with a single bank and ditch, but weak in its position. It stands in a belt of woodland that runs East and West; to the North and South cultivation prevents any conjecture as to ancient habitation that may have surrounded this camp, and on the North side the land has been grubbed, as well as ploughed—according to the evidence of Thos. Aldwell's " Mappe of Cranburne Chace." The camp site is about half a mile West of the road leading from Handley to Bower Chalke, standing on ground that slopes gently from North to South, about 490 feet above the sea, on the Dorset side of the " Shire Rack " (the boundary between Dorset and Wilts). The whole of its area and entrenchment are overgrown with hazel underwood interspersed with oak trees. The entrance on the South-Eastern side appears to be original. From the scale of the entrenchment we may be sure that its purpose was defensive. The planted area shows neither humps nor hollows, and I have not found potsherds upturned in the rabbit scrapes. In size, this camp compares with those on Penbury Knoll, Damerham Knoll, and Bussey Stool Park (named " Hook's Wood " in Warne's " Ancient Dorset "), but its entrenchment is greater than those of the first two, and less than that of the last. In Thos. Aldwell's " Mappe of Cranburne Chace," 1618, Mistleberry is named Mapleberry. This camp was overlooked by Warne. About a quarter of a mile North-West of the camp, a wasted ditch, of the Grim's Ditch type, may be traced near Stone-down Wood, which eventually leads to the British settlement at the back of Middle Chase Farm.

[1] See " The Early Wars of Wessex," by Albany F. Major and Charles W. Whistler.

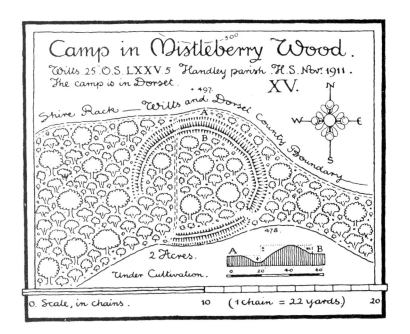

Camp in Mistleberry Wood.

Wilts 25" O.S. LXXV. 5 Handley parish. H.S. Nov. 1911.

The camp is in Dorset.

XV.

Shire Rack — Wilts and Dorset County Boundary

+ 497.

-500-

A

B

478.

2 Acres.

Under Cultivation.

A   B

0   10'   40'   60'

0. Scale, in chains.   10   (1 chain = 22 yards.)   20

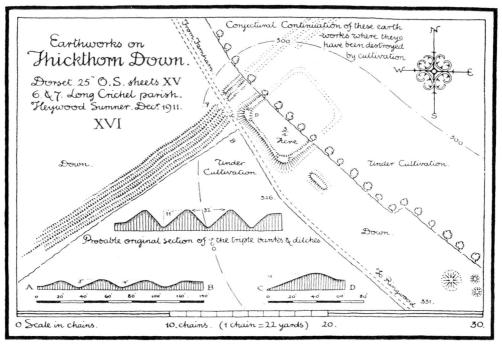

Earthworks on
Thickthorn Down.

Dorset. 25″ O.S. sheets XV
6. & 7. Long Crichel parish.
Heywood Sumner. Dec.r 1911.

XVI

Conjectural Continuation of these earth
works where they
have been destroyed
by cultivation

300

Down.

Under
Cultivation

¾
Acre

Under Cultivation.

326.

Down.

Probable original section of the triple banks & ditches

To Ringwood

331.

O Scale in chains.   10. chains.   (1 chain = 22 yards)   20.   30.

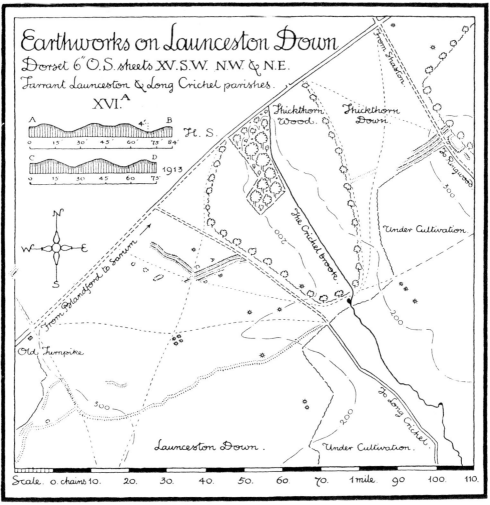

Earthworks on Launceston Down

Dorset 6″ O.S. sheets XV. S.W. N.W. & N.E.
Tarrant Launceston & Long Crichel parishes.

XVI.A

H. S.

1913

From Shaston

Thickthorn
Wood.

Thickthorn
Down.

To Ringwood

300

Under Cultivation

The Crichel brook.

200

From Blandford to Sarum

Old Turnpike

300

200

To Long Crichel.

Launceston Down.

Under Cultivation.

Scale. o. chains 10.   20.   30.   40.   50.   60.   70.   1 mile   90   100.   110.

EARTHWORKS ON THICKTHORN DOWN

THE site of these earthworks is beside the long uphill road from Ringwood to Shaftesbury, on the ridge that divides the Gussage valley (to the East) from the Crichel valley (to the West). They denote occupation during successive prehistoric periods, for here may be seen: A long barrow, three round barrows, a triple row of banks and ditches, and the fragment of a square or oblong camp, all standing within a distance of 400 yards. No other camp on Cranborne Chase quite compares with this fragment on Thickthorn Down. It is defended by a single bank and ditch. The bank is well preserved on the South-Western side, and rises to

NOTE REFERRING TO THE DIAGRAM.

Each square of excavated ditch soil is supposed to be piled upon the corresponding square of bank; thus it will be seen that the corner square of bank has to take three squares of excavated soil.

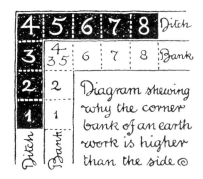

Diagram shewing why the corner bank of an earthwork is higher than the side

12 feet from the bottom of the ditch. The other two sides are curtailed, wasted, and worn by cart traffic. The ditch is broad and shallow, and cut into by the road at the Western corner. There is no entrance discernible in the fragment that remains. The bank is very much raised at both corners. This is not owing to defensive requirement, but to the nature of things, i.e., earthworking. The corner of a rectangular earthwork bank must take three times the amount of soil to what the side bank takes (see diagram). North-West, beyond this camp fragment, the road passes on through a row of banks and ditches, triple on the South-Western side, double on the North-East, similar to those on Gussage Down and Launceston Down. Both the camp and the ditches are apparently defensive, but their abrupt termination (under cultivation) on the North-Eastern side of the down leaves their extent and plan uncertain.

It will be noticed that on the North-Eastern side of the road the triple row of banks and ditches become double. The superficial appearance of these earthworks suggests that the missing third bank may have been used to fill in the missing ditch by the camp makers—who I conjecture were of later period than the makers of the bank and ditch entrenchments—so as to provide an outer berm platform on which the camp defenders could fight, with the camp rampart behind them as their second line of defence.

In considering the defensive value of this triple row of banks and ditches that now appear both low and shallow, we may note that the ditches have probably silted up about 4 feet in depth (see sections cut in Grim's Ditch, Plan XXXIV), and the banks have been wasted to a corresponding extent. When first thrown up the top of each of these triple banks was probably about 12 feet above the bottom of the ditch, and 32 feet across from bank top to bank top (see probable original section in the plan). Such earthworks would prove an effective barrier to stop a rush of raiding horsemen.

The enemy against whom these defences were raised seems to have been coming from the North-West. It will be seen by referring to the plan that the camp is placed in the centre of the ridge end bounded by the 300-foot contour line, and the triple row of banks and ditches also appear to have been originally made so as to protect this hill ridge. The dotted lines shown in the plan describe a conjectural continuation of these fragmentary earthworks on the land under cultivation, where now they are wholly effaced. Only careful excavation could yield an answer to the questions here suggested; but the site is comparatively small, and is free from tree-planting, accordingly excavation is not an impracticable counsel of perfection.

The ditches of Thickthorn Down are quite effaced by cultivation in the valley of the Crichel brook, but they reappear on Launceston Down, on the West side of the road leading from Long Crichel to the Blandford and Sarum high road. Here are two ditches and a triple row of banks, but smaller than those on Thickthorn Down, their all-over measurement of 84 feet comparing with 120 feet at the latter site. These ditches die out beyond the intersection of two down tracks (see plan). Just before this, a single ditch, between two well-preserved banks, branches off towards Tarrant Hinton Down, but dies away before reaching the high road.

There is another ditch that rambles at length across Launceston Down (see plan), the connection of which is not apparent now. On the top of the Down, near the old turnpike, it has curious branches and multiplications. The Down surface is here scored by tracks, and thereby the trace of these ditches is made uncertain; but it seems that the Ordnance Survey has omitted two of these branches, while the " Tumulus " marked on the Northern side of the road appears to be an isolated fragment of banks and ditches similar to those on Launceston Down (see plan).

The all-over measurements of these entrenchments are fairly similar —34 to 35 feet—and they appear to belong to the same period of prehistoric landmarks. Boundaries and cattle stops are the purposes which I am inclined to think they served. The occasional increase of breadth in the middle bank, where the ditches are multiplied, is a point to be noted, both here, on Gussage Down, and, to a lesser degree, on Thickthorn Down. Perhaps random setting out may be the cause of this feature. The planning of these entrenchments does not suggest deliberate intention in every detail, such as is suggested, for example, by the entrenchments of Lydsbury Rings on Hod Hill, or of Soldier's Ring near Damerham.

I have not found any potsherds on the rabbit scrapes or mole hills on Launceston Down, but suspect British habitation on this site.

ENTRENCHED ENCLOSURES, PROBABLY FOR PASTORAL USAGE

THE ROMANO-BRITISH ENCLOSURE ON ROCKBOURNE DOWN

THIS large enclosure has recently been discovered by the writer in the course of this survey, and excavation (by kind permission of the Earl of Shaftesbury) has shown that it belonged to the period of the Roman occupation. Here, unlike the sites at Rotherley, Woodcuts, and Woodyates, there are no signs of earlier British entrenchments, or of the adaptation of an open British village site, to the later requirements of the Romanized Britons. The planning and execution of the entrenched enclosure, and of the principal habitations connected with it, are essentially Roman. Three well-built hypocausts were found. One belonging to a small dwelling-house, 18 feet by 10 feet, with a bakehouse adjoining, 10 feet by 7 feet; while a third, 12 feet square—which was placed at the corner of a small, ditched quadrangle, within which were evidences of habitation—served apparently for corn-drying, malting, and cooking purposes. The coins found date from A.D. 253 to 375.

The low, broad entrenchments which surround the enclosure are very remarkable. They have no parallel in this district. Excavation proved the ditches to be wide, very evenly cut, and the inner ditch to be wider and deeper than the outer. Between these ditches there intervened a space of 14 feet, on which was dumped the earth from the digging of the ditches, thus making a central bank. The scarps of the ditches were much steeper on the sides of this central bank than were the counter-scarps on the area side and on the outside. Post holes, for a stockade, were found along the bank scarp of the inner ditch, with intermediate holes on the bank. Apparently the cattle stockade was made in a zig-zag line in order to lessen wind pressure. Remains of wood—either burnt or decayed—were found in the post holes.

A full account of these excavations, with details, will be given shortly.

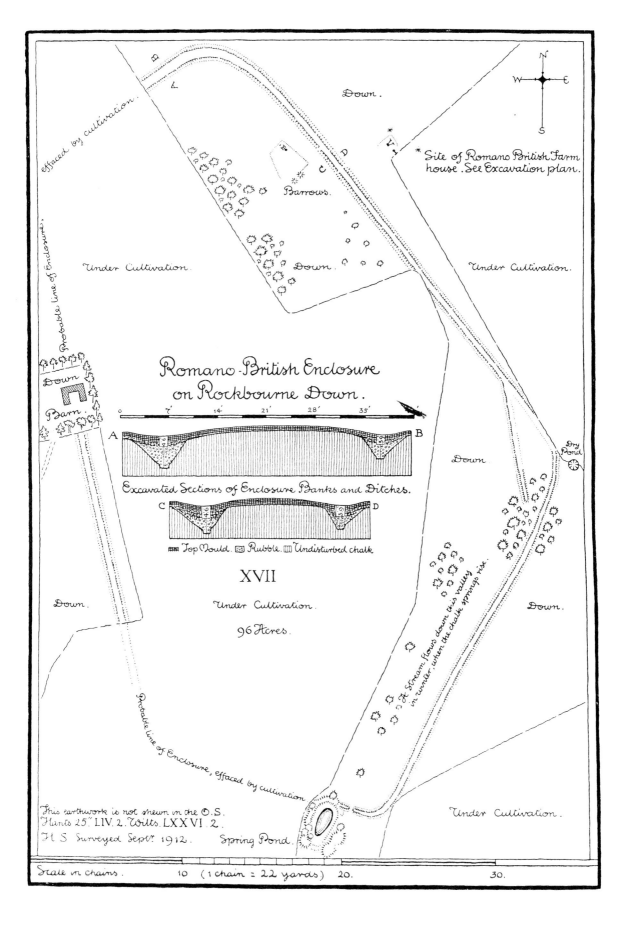

Down.

* Site of Romano British Farm house. See Excavation plan.

Barrows.

Under Cultivation.

Under Cultivation.

Effaced by cultivation.

Probable line of Enclosure.

Down.

Barn.

Down.

Down.

Under Cultivation.

Down.

Dry Pond.

Romano-British Enclosure on Rockbourne Down.

0   7′   14′   21′   28′   35′

A   B

Excavated Sections of Enclosure Banks and Ditches.

C   D

▦ Top Mould. ▨ Rubble. ▥ Undisturbed chalk.

XVII

Under Cultivation.

96 Acres.

A stream flows down this valley in winter, when the chalk springs rise.

Probable line of Enclosure, effaced by cultivation.

Down.

Under Cultivation.

This earthwork is not shewn in the O.S.
Hants 25″ LIV. 2. Wilts. LXXVI. 2.
H S Surveyed Sept. 1912.   Spring Pond.

Scale in chains.   10   (1 chain = 22 yards)   20.   30.

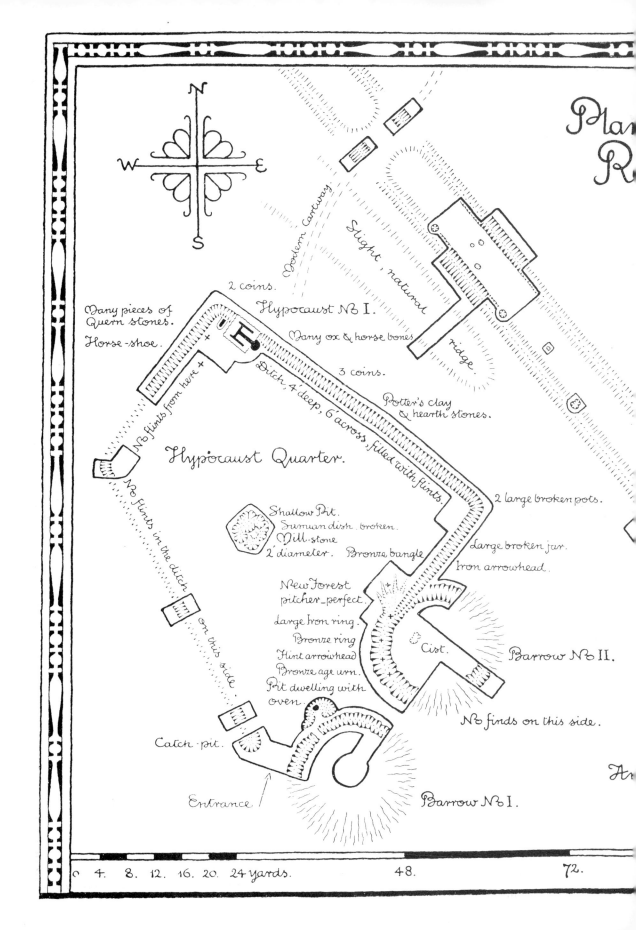

Plan
R

N
W E
S

Modern Cartway.

Slight, natural ridge.

2 coins.

Hypocaust Nº I.

Many pieces of Quern stones.

Horse-shoe.

No flints from here +

Many ox & horse bones

3 coins.

Potter's clay & hearth stones.

Ditch, 4 deep, 6 across, filled with flints.

Hypocaust Quarter.

No flints in the ditch on this side.

2 large broken pots.

Shallow Pit.
Samian dish broken.
Mill-stone
2 diameter.   Bronze bangle

Large broken jar.

Iron arrowhead.

New Forest
pitcher–perfect.

Large Iron ring.

Bronze ring

Flint arrowhead

Bronze age urn.

Pit dwelling with oven.

Cist.

Barrow Nº II.

No finds on this side.

Catch-pit.

Entrance

Barrow Nº I.

An

f Excavations on
bourne Down
S. 1911—1913

Drain Ditch.

2 coins.

Hypocaust No II.

Purbeck marble bowl.

Hypocaust No III.

Slight ridges & hollows cross the down here, trending in the direction of the Hypocausts. Trial trenches cut across them in three places showed that they were natural. The down level falls gently from N.W. to S.E.

1 coin. Iron arrow-head.

flints

Flint Causeway across the Entrenchments with post-holes for a gate on the inner side.

XVII.A.

Notes. The excavated portions are shewn in this plan surrounded by a firm outline.

Enclosure Entrenchments.

The Enclosure Entrenchments appear on the down surface merely as slight continuous depressions where the grass grows green, on either side of a broad, low, continuous bank on which the grass grows grey-green.

The ditch of Barrow No II was filled with collected flints where the four +'s are marked. Elsewhere, the ditches of both Barrows were filled with mould above, & chalk rubble below.

Enclosure.

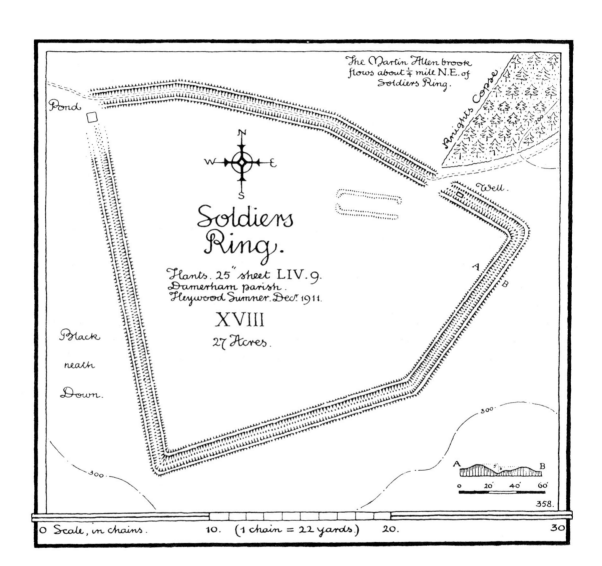

The Martin Allen brook
flows about ¼ mile N.E. of
Soldiers Ring.

Knights Copse

Pond

Well.

Soldiers
Ring.

Hants. 25" sheet LIV.9.
Damerham parish.
Heywood Sumner. Dec.ʳ 1911.

XVIII

27 Acres.

Black

neath

Down.

A
B

300

300

A   B
0   20′ 40′ 60′

358.

SOLDIER'S RING

THE entrenchments that surround Soldier's Ring are different from any that exist on Cranborne Chase. The site includes two little valleys that run up to the South-West from the Martin Allen Valley, and it seems probable that the springs would have actually risen within the enclosure when it was made. It is commanded on three sides by rising ground, and is perversely named, for the camp cannot be supposed to have served any military purpose, while its slight entrenchments surround a pentagonal, and not a circular, area. Probably it served as a cattle enclosure, and the site was well suited for pastoral requirements. The run of the banks is extraordinarily even, with the exception of the usual rise at the corners (see p. 35 for the reason of this). The precision of the entrenchments, both in setting out and in execution, is specially notable, also the low inside bank that rises as sort of shoulder on the counter-scarp of the outer bank (see plan). The entrance seems to have been at the lower Eastern side of the enclosure. Near the entrance, within the area, there are two low parallel banks partly enclosing an oblong space (see plan), the purpose of which might be discovered by excavation. On the North-Eastern side have been found a few sherds of wheel-turned pottery, but there is no place on this site that, on the surface, specially suggests habitation. The precise execution and planning of these entrenchments indicate the work of the Romanized Britons. I do not think that the pre-Roman Briton could have set out these exact alignments of bank and ditch, or have enclosed so large an area without making some variation in the scale and direction of the earthworks.

CHICKENGROVE AND MARLEYCOMBE HILL

DUE South of Bower Chalke rises Marleycombe Hill. Approached from the village, it appears as a great promontory of down land, jutting into the sheltered valley, with a white road winding round its steep Eastern scarp. " The hill-side is the valley wall." But if approached from Chickengrove Bottom, which lies about half a mile South of its summit, Marleycombe Hill merely appears as the termination of a high, cultivated upland, the climax of the gradual rise in the land that begins about three miles off, farther South, on Martin Down. Here, in a rough pasture,

39

partly overgrown with furze, on the Southern side of the Oxdrove Ridgeway, may be found the earthworks as shown in the plan. Probably they belong to a defensible cattle enclosure, but they are mere wasted remains of a camp that has for the most part been effaced by cultivation. No part of the entrenchments can be measured as a representative section.

According to Sir R. C. Hoare's survey of this district in 1810-12, a British village could then be traced between Marleycombe Hill and the Ridgeway. Since then the plough has been at work, and now there are no traces of primitive habitation, except tell-tale cooking flints, which may be picked up on the arable.

Chickengrove Bottom is a dry combe that runs up into the Ridgeway Hill from Martin Drove End. The track to Broadchalke follows the lower end of the Bottom, with Vernditch Chase to the West, where Grim's Ditch pursues its twisting course, and the desolate site of Vernditch Lodge to the East is marked by a beautiful group of trees. After a few hundred yards of grass track, through an open woodland valley, the road again becomes metalled, turns up the hill to the right, and mounts the cultivated hillside that flanks the Oxdrove Ridgeway on the South. The earthworks above Chickengrove Bottom may be reached either by taking this road, and then following the Ridgeway track on the top, to the left, for half a mile, or, preferably, by rambling up the mossy combe of Chickengrove Bottom to its head; then a gate to the right leads through rough coppice into a pasture, reverted, furze-covered, humped with mole hills, littered with rabbit scrapes (whereon sherds of British pottery may be found), and here are the earthworks that once belonged to a British settlement.

Now only a large, irregular entrenchment and cultivation banks can be traced. The incurving banks of the South entrance are fairly well preserved, but for the most part, both the bank and the ditch that surrounded the enclosure are spread and flattened by cultivation. On the Eastern side of the entrance there is a curious sort of projecting bastion. From what now remains of this entrenched enclosure it suggests comparison with those on Tarrant Hinton Down and South Tarrant Hinton Down, but it is much larger than either of these. It is entirely different in type from the purely pastoral enclosure on Rockbourne Down, and I conjecture that it was made by the Britons in pre-Roman times. The position is elevated, 600 feet above the sea, on a gentle slope facing South-East. The earthworks suggest both pastoral and defensive purposes, but defence does not appear as the main purpose of the enclosure.

This settlement must have been a large one, for (according to Sir

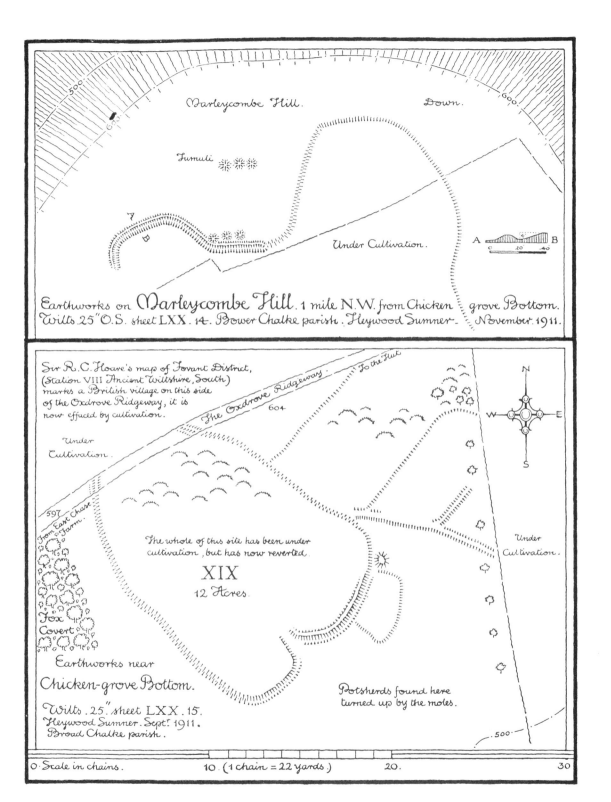

Marleycombe Hill.   Down.

Tumuli

Under Cultivation.

A B
0 10' 40

Earthworks on Marleycombe Hill. 1 mile N.W. from Chicken-grove Bottom.
Wilts. 25" O.S. sheet LXX. 14. Bower Chalke parish. Heywood Sumner. November. 1911.

Sir R.C. Hoare's map of Fovant District,
(Station VIII Ancient Wiltshire, South)
marks a British village on this side
of the Oxdrove Ridgeway, it is
now effaced by cultivation.

The Oxdrove Ridgeway.   To the Flat

604

Under
Cultivation.

597

From East Chase Farm.

Under
Cultivation.

The whole of this site has been under
cultivation, but has now reverted.

XIX
12 Acres.

Fox
Covert

Earthworks near

Chicken-grove Bottom.

Wilts. 25" sheet LXX. 15.
Heywood Sumner. Sept! 1911.
Broad Chalke parish.

Potsherds found here
turned up by the moles.

500

N
W   E
S

0. Scale in chains.   10. (1 chain = 22 yards.)   20.   30

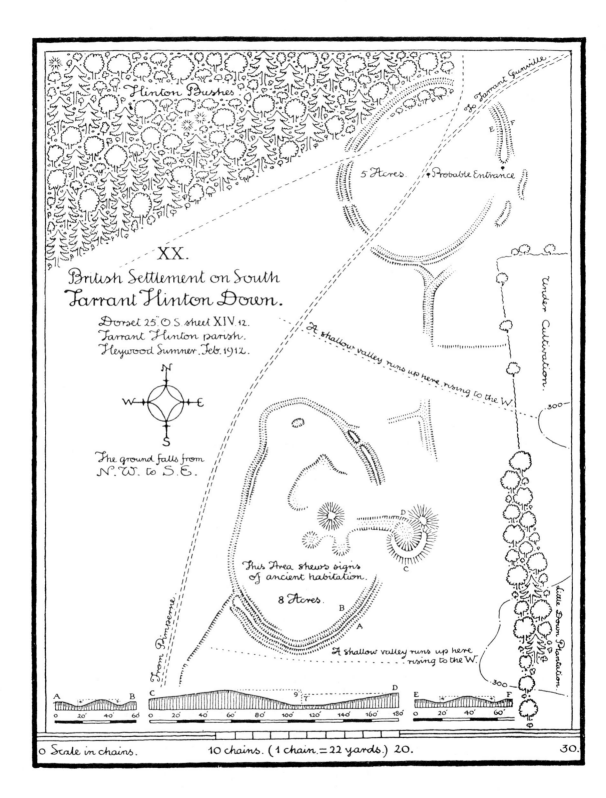

Hinton Bushes

To Tarrant Gunville

E F

5 Acres   Probable Entrance

XX.

British Settlement on South
Tarrant Hinton Down.

Dorset 25. O.S. sheet XIV. 12.
Tarrant Hinton parish.
Heywood Sumner. Feb. 1912.

N
W   E
S

The ground falls from
N.W. to S.E.

Under Cultivation.

A shallow valley runs up here, rising to the W.

300

D

C

This Area shews signs
of ancient habitation.

8 Acres.

B

A

A shallow valley runs up here
rising to the W.

Little Down Plantation.

300

From Pimperne.

A   B   C   9 7   D   E   4   F
0   20' 40' 60'   0   20' 40' 60' 80' 100' 120' 140' 160' 180'   0   20' 40' 60'

R. C. Hoare) the British village beyond the Ridgeway would have connected it with the enclosure on Marleycombe Hill, of which only a fragment remains. Thus the two enclosures lie on either side of the upland,
like panniers on its broad back, with the British village site on the
Ridgeway plateau connecting them.

It should be noted that the banks and ditches of the Chickengrove
enclosure rise and fall in two places right across the Oxdrove Ridgeway,
showing that the present way does not keep the line of the ancient
trackway that presumably followed this ridge.

There is a glorious view here—Vernditch in the foreground, Pentridge and Damerham Knoll in the middle distance, the forest beyond,
and on, till in the far distance, the Isle of Wight fades into the clouds of
the horizon.

SOUTH TARRANT HINTON DOWN

THE settlement on South Tarrant Hinton Down is specially interesting.
Here are two oval enclosures, surrounded by slight entrenchments that
are separated from each other by a shallow down valley, in which presumably the water rose when these enclosures were made. The upper
enclosure shows no superficial signs of ancient habitation, but there is a
sunken way leading down to the little valley that suggests cattle usage.
It should be noted that outside the entrance on the Eastern side stand the
wasted remains of two detached banks that appear to be defences covering the opening, and that the Northern bank of the enclosure (the
Southern bank has been destroyed) widens into a pear shape at the
entrance—a form that often occurs at camp entrances. The all-over
measurements of the entrenchments and the situation show, however, that
this camp can never have been of much account, and I am inclined to
regard it as a pastoral enclosure with slight defences.

The lower enclosure is the larger of the two, and the area is covered
with humps and hollows and dense green grass that suggest habitation.
The entrance is on the South-Eastern side. On the North-Eastern side
there is a semicircular depression, strongly banked, and approached from
the area by a sunken way. This compares with somewhat similar earthwork forms on Hambledon Hill, Gussage Down, Blandford Race-Down,
Tarrant Hinton Down (near Eastbury), Chettle Down, and at Buzbury
Rings. Their purpose might be discovered by excavation. Meantime,
conjecture may be allowed, even to the superficial surveyor; and accord-

ingly I suggest that these semicircular, banked depressions may have been constructed as cattle shelters—to screen the herds from the wind. These down sites are absolutely exposed to the elements; somehow, the cattle must have been sheltered in rough weather; and I have learnt from frequent experience how effective is the wind screen afforded by such banks —even in their present wasted condition. The requirement, and such fulfilment, seem possibly to fit. Sunken pits were dug by the Britons for their dwellings on these open sites, and their example was followed by the Romanized Britons when they made their hypocausts on these downs in the early centuries of our era. So, embanked and sunken areas for cattle shelters would have been a natural and traditional extension of comfort from man's experience to beast's necessity.

There is another puzzle suggested by this enclosure, in the duplication of the single entrenchment, which for the most part surrounds it, at the lowest side of the site.

A ditch, between double banks of the Grim's Ditch type, starts from this low South-Western side, and can be traced for some distance over the hill towards Pimperne.

TARRANT HINTON DOWN (NEAR EASTBURY)

THIS is a site where many traces of British occupation remain. The cattle enclosure is large, and there are many wasted earthworks adjoining. The Down appears to have been under cultivation " once upon a time," hence the wasted condition of the earthworks. Boundary banks and ditches approach the site from Launceston Down and from the Tarrant Valley, suggesting that this was a centre from which boundaries radiated. There is a curious horseshoe form of sunken entrenchment within the area, as to the purpose of which I have made a conjecture in my previous notes on South Tarrant Hinton Down. The situation of these entrenchments, and their apparent scale, do not indicate defence as a primary requirement. We may assume that the water came out at the bottom of the enclosure when it was made, and thus it would have been well adapted for pastoral purposes.

The beauty of the hedgerows about here is specially notable. Ancient yews, hollies, thorns, and ivy trees, spindle, sloe, and dog-wood bushes line the Down limits with a freedom of growth that suggests the survival of primitive nature. Perhaps these wild hedgerows may be relics of the ancient forestry of the chase.

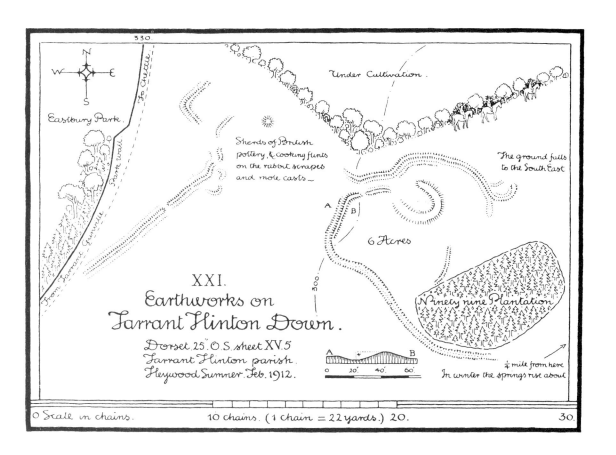

330

N
W ◇ E
S

To Chettle

Eastbury Park.

Park Wall.

From Tarrant Gunville.

Under Cultivation.

Sherds of British
pottery, & cooking flints
on the rabbit scrapes
and mole casts—

The ground falls
to the South East

A
B

6 Acres

300.

Ninety nine Plantation

XXI.
Earthworks on
Tarrant Hinton Down.
Dorset 25". O.S. sheet XV.5
Tarrant Hinton parish.
Heywood Sumner. Feb. 1912.

A   B
0   20'   40'   60'

¼ mile from here
In winter the springs rise about

0 Scale in chains.   10 chains. (1 chain = 22 yards.) 20.   30.

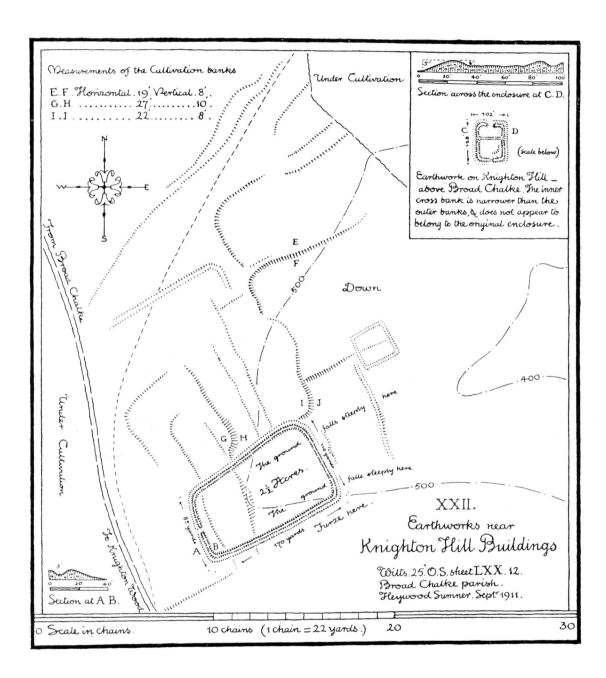

Measurements of the Cultivation banks

E.F. Horizontal . 19' . Vertical . 8' .
G.H. .... 27' ....10'
I.J. .... 22' .... 8'

Under Cultivation

Section across the enclosure at C.D.

Earthwork on Knighton Hill _
above Broad Chalke. The inner
cross bank is narrower than the
outer banks, & does not appear to
belong to the original enclosure.

C ⊢ 102' ⊣ D
(scale below)

From Broad Chalke

Under Cultivation

To Knighton Wood

Down

E
F
500

I J
falls steeply here

G H
The ground
2½ Acres
The ground
falls steeply here

85 yards

A B
170 yards   Furze here

400

500

XXII.
Earthworks near
Knighton Hill Buildings
Wilts. 25" O.S. sheet LXX. 12.
Broad Chalke parish.
Heywood Sumner. Sept.r 1911.

Section at A B.

0 Scale in chains.   10 chains (1 chain = 22 yards.)   20   30

EARTHWORKS NEAR KNIGHTON HILL BUILDINGS

THE earthworks near Knighton Hill Buildings lie about half a mile North of the Oxdrove Ridgeway and of Knighton Wood. They stretch along the Southern side of a lateral combe that runs up from the Ebble Valley into the great chalk escarpment of the Ridgeway.

The enclosure is well defined, very precisely entrenched, and is surrounded by a low bank with a shallow ditch on the outside, which cuts the scarps of three cultivation banks—showing that these were in existence, and that possibly the land had reverted when the enclosure was made.[1] It lies at the top, West end of the combe, and extends half-way down. I have not found any sherds in the rabbit scrapes or in the mole-hills, nor any signs of habitation, such as cooking flints. Cultivation banks abound on these hill-sides, but the probable sites of habitation on the hill-tops are now beneath the ploughed land.

The entrance of the enclosure is in the middle of the Western side, but is not shown on the Ordnance Survey Map. The additional height of the bank at the corners may be noted—for explanation of this feature see p. 35. This enclosure was probably used as a pastoral fold, and it compares with the earthwork at Church Bottom on Prescombe Down, near Ebbesbourne Wake, and that on Fifield Down. Sir R. C. Hoare seems to refer to this enclosure on p. 247, "Ancient Wiltshire, South," but if so, he places it wrongly in a side combe of Church Bottom instead of Croucheston Bottom. Certainly his description fits the earthwork that we are surveying, and certainly no traces of such an earthwork are now to be seen on the downland site which he indicates.

The cultivation banks or linchets that line the hill-sides from Clearbury, past Little Yews, Homington, Coombe Bissett, Knighton, Broad Chalke, and on up the valley of the Ebble, tell of peace and of the persistent cultivator. They remind us of livelihood. The hill-top camps of defence and refuge are so concentrated, so effective, and so permanent that we are apt to think of earthworks merely in terms of warfare, but we have to account for all the dim centuries since man emerged from the hunter stage, and became first a herdsman and then also a cultivator, and the British settlements, the cultivation banks, and the pastoral enclosures

---

[1] *Cf.* "The Problem of Ancient Cultivations," by Herbert S. Toms, in "The Antiquary," November 1911. Fig. 2 shows "A valley entrenchment built over ancient cultivations in Eastwich Bottom, near Brighton." Also "Notes on Some Surveys of Valley Entrenchments in the Piddletrenthide District" ("Dorset Field Club Proceedings," vol. xxxiii, by the same author).

surely suggest long periods when the prehistoric inhabitants of these hill-tops looked forward to reaping where they had sown, and to folding their sheep and cattle in pastoral rather than in defensive enclosures. They imply a life of agriculture and of stock-keeping, not a life of marauding. Even the boundary ditches are testimonies to a widespread and tolerable order in life during the period that merges from the pre-historic to the historic—probably we may say to Anglo-Saxon times. Then, the abandonment of the old hill-top British sites in favour of the valleys does suggest a period of catastrophe, and then we can imagine that the camps were sought in vain for refuge and defence, that the pastoral enclosures were empty, and that the down grass was encroaching on the cultivated terraces that had been wrought so laboriously on the sides of the hills.

Below the enclosure (on the North side of the combe) is another and a smaller earthwork, with a narrow bank and outside ditch, so narrow as to suggest a modern origin, but the dimensions of this small enclosure are curiously similar to those of the earthwork on Knighton Hill, subsequently described and shown in the inset plan—namely, 47 paces by 39 paces the former, 46 paces by 33 paces the latter.

On Knighton Hill, above Broad Chalke, there is a small rectangular earthwork, with a broad bank and outside ditch. The area is divided in the middle by a narrow bank and ditch which may be modern; it seems to be an addition to the original enclosure, as its slight ditch cuts the outside banks on both sides, and the upturned earth makes humps on the outside banks at the lines of intersection. The area is uneven, in mounds and hollows. The moles work freely in the silt of the ditch, but turn up no potsherds.

This earthwork was probably made for pastoral purposes. It is mentioned by Sir R. C. Hoare in "Ancient Wiltshire, South."

CHETTLE DOWN

THE settlement on Chettle Down is remarkable for the large number of cultivation, or enclosure, banks that cover the down surface. The area of habitation was apparently at the Northern side of the down, above the small oval, wide-banked enclosure. Here, beside two mounds of unusual form, sherds of British pottery may be found upturned wherever the moles have been working. A sunken way leading from this oval enclosure, down

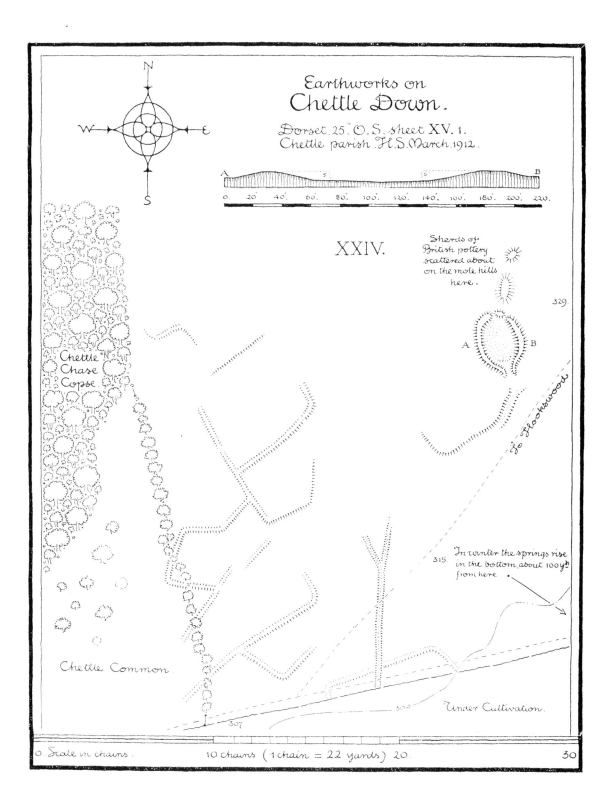

Earthworks on
Chettle Down.

Dorset. 25". O.S. sheet XV. 1.
Chettle parish. H.S. March. 1912.

XXIV.

Sherds of
British pottery
scattered about
on the mole hills
here.

329

Chettle
Chase
Copse

To Hookswood

315

In winter the springs rise
in the bottom, about 100 y⁴ˢ
from here.

Chettle Common.

Under Cultivation.

307

300

0. Scale in chains.   10 chains (1 chain = 22 yards) 20.

30

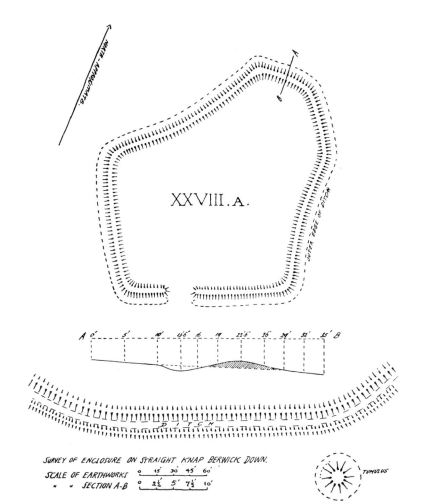

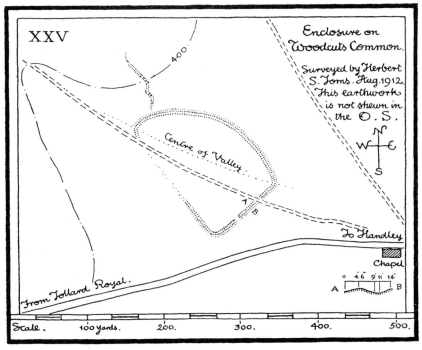

XXVIII.A.

NORTH APPROXIMATE

OUTER EDGE OF DITCH

A 0' 5' 10' 156' 16' 19' 22'6 16' 24' 32' 35' B

DITCH

SURVEY OF ENCLOSURE ON STRAIGHT KNAP BERWICK DOWN.

SCALE OF EARTHWORKS   0   15 30 45 60'

"   "   SECTION A-B   0 2½ 5' 7½ 10'

TUMULUS

XXV

400

Centre of Valley

Enclosure on
Woodcuts Common.

Surveyed by Herbert
S. Toms. Aug. 1912.
This earthwork
is not shewn in
the O.S.

N
W E
S

A B

To Handley

Chapel

0 4.6 9 11 14'
A   B

From Tollard Royal.

Scale.   100 yards.   200.   300.   400.   500.

the hill towards the South, suggests a cattle-way to the water—which now rises at the bottom of the down in wet seasons. It compares with similar sunken ways on South Tarrant Hinton Down, and on Blandford Race Down, but in the latter instance, the sunken way, though leading to the water, is not, now, plainly connected with an enclosure. The oval shaped, sunken, and entrenched enclosure, compares in plan with that given in Warne's " Ancient Dorset " of an earthwork (now partially destroyed) within Buzbury Rings. This site appears to have been occupied by the pre-Roman Britons for pastoral purposes.

WOODCUTS COMMON

THIS enclosure has been recently surveyed, and completely traced by Mr. Herbert S. Toms. Only the Southern corner of the earthwork had been shown in previous surveys. It seems to belong to the type of valley pastoral enclosures,[1] but its entrenchment is less than that of any other earthwork of a similar type on Cranborne Chase.[2]

CHURCH BOTTOM, PRESCOMBE DOWN

THIS enclosure compares with those near Knighton Hill Buildings and on Fifield Down, in precision of setting out, and in scale of entrenchment. The Southern entrance seems to have been the original one. That on the Northern side has probably been made by cattle traffic up and down the Bottom. The Western bank has been partially destroyed, otherwise this earthen fold is well-preserved, and is a beautiful example of ancient, peaceful, pastoral care. (See inset, Plan XXXVI.)

BERWICK DOWN

See p. 74, *infra*.

---

[1] Mr. Herbert S. Toms makes a provisional classification of these earthworks as follows: " A. Valley-head enclosures. B. Valley-side enclosures. C. Valley proper enclosures—*i.e.*, so designed as to enclose not only the valley floor, but portions of each side of the valley as well." Vol. xxxiii, " Dorset Field Club Proceedings."
[2] A full account of this enclosure is given by Mr. Herbert S. Toms in " The Antiquary," July 1913.

EARTHWORKS OF EXCEPTIONAL CHARACTER

KNOWLTON

IT is doubtful whether Knowlton was within the ancient outbounds of Cranborne Chase. The place-names of the perambulation are dubious here. Even the local knowledge of Dr. Wake Smart was baffled in their identification.[1] But it is so near to the limits of our district that a few yards may be conceded, and we may take the benefit of the doubt.

For it is a benefit. It enables us to consider a most remarkable site—remarkable in this respect. Nowhere on Cranborne Chase, excepting, perhaps, the disk barrows near Woodyates, do we find any earthwork expression of what is supposed to be prehistoric formular religion. Circles, either marked by stones, or wrought in earth, are the signs of the unknown religion of our forefathers. Here, at Knowlton, surrounded by barrows, we have four circular earthworks, only one of which is still perfect, the others having been destroyed by cultivation. But from the remnant that remains, and from the situation, we cannot suppose that purposes of defence or of cattle enclosure were the motives of the makers of these rings. The two (apparently) original entrances of the one perfect remaining circle are opposite each other. The wide ditch is on the inside. The bank is unusually broad and precise in its circling enclosure. There is no similarity in its construction to that of any of the other earthworks on Cranborne Chase. Warne[2] attributed these Rings at Knowlton to the Druids. But the Druids are so incapable of proof by means of excavation, that it is now usual to avoid them, and to take refuge in " some form of " solar religion." The distinction may represent a difference. It is difficult to appreciate. It is possible, however, to appreciate the absolute difference in design between these earthworks and the others included in our survey. These that we are considering express a precision, apart from defensive or pastoral purposes, that embodies a new motive in construction, and the

---

[1] " A Chronicle of Cranborne." See Appendix.   [2] " Ancient Dorset."

46

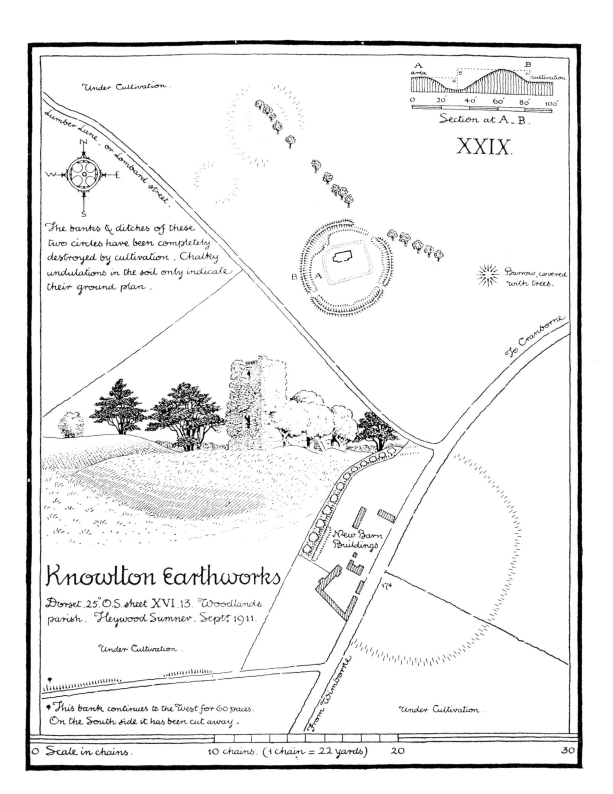

Under Cultivation.

Lumber Lane, or Lombard Street.

N

W E

S

The banks & ditches of these
two circles have been completely
destroyed by cultivation. Chalky
undulations in the soil only indicate
their ground plan.

A   area

B   cultivation

0   20'   40'   60'   80'   100'

Section at A–B.

XXIX.

B   A

Barrow, covered
with trees.

To Cranborne

New Barn
Buildings

.174

From Wimborne

Knowlton Earthworks

Dorset. 25" O.S. sheet XVI.13. Woodlands
parish. Heywood Sumner. Sept? 1911.

Under Cultivation.

Under Cultivation.

• This bank continues to the West for 60 paces.
On the South side it has been cut away.

0   Scale in chains.   10 chains. (1 chain = 22 yards)   20   30

vague knowledge that we have of the Druids from Caesar, tells us that they introduced precise forms of religious observance and of government. And circles seem to have played a part in their formular expression. When we pass the wine with the Deisal turn, we perhaps unconsciously acknowledge Druidical belief. And so, also, when we look askant at the Tuaphol turn or Widdershins. There seems to be a different directing mind expressed in the making of these ringed earthworks, with their ditches on the inside, from that of the British camps and settlements and cattle enclosures.

In Hadrian Allcroft's "Earthwork of England," there are two illuminating chapters on Miscellaneous Earthworks, in the second of which he discusses these Rings at Knowlton, and shows how they compare in certain particulars with those at Thornborough Moor, near Ripon, and with the camp at Figsbury Ring, near Salisbury. To this chapter I desire to refer my reader.

Two of the destroyed Rings at Knowlton are practically effaced. They are now merely circular undulations on arable land. But the largest of the four still suggests its original construction in one place—at the back of New Barn Buildings. Here, for a short space, both the outer bank and the inside ditch of this Ring are partially discernible.

Knowlton has given its name to a Hundred—showing that it was a place of special significance in Anglo-Saxon times. Subsequently, probably in the fourteenth century, a little stone church was built within the area of the circle that still remains, and it is further surrounded by a low, oblong bank. Particulars of its ecclesiastical vicissitudes may be found in Hutchins's "History of Dorset," vol. iii. The site of this ruined Christian church, standing within an earthen circle that seems to belong to the unknown religion of the early Britons, and guarded without by a row of ancient yew trees, is indeed most beautiful, and it marks the strange changes and chances that have happened during the lapse of 2,000 years.

Reference to the plan of Knowlton will show that these circles are surrounded by barrows—like Stonehenge and like Avebury. But it must also be noted that this site does not now appear as the barrow centre of Cranborne Chase—as Stonehenge is the barrow centre of Salisbury Plain. That distinction belongs to Oakley Down below Pentridge, near Worbarrow.

ACKLING DYKE AND BARROWS NEAR OAKLEY DOWN

THE site of this group of earthen monuments to the mighty dead of antiquity must once have been specially reverenced. Nowhere on Cranborne Chase will you find, close together, such various and imposing barrows—long, bowl, bell, and disk. They are all to be found here, and they represent the finest of their kind. This place is a British Campo Santo between the derelict Ackling Dyke, on one side, along which the Roman legions tramped from Durnovaria to Sorbiodunum in the early centuries of our era, and the white highway on the other side, along which motors now rush at top speed over the solitary downs. Both roads in their straight courses have cut into the barrow circles, and both thus express a silent disregard for departed glory.

Dr. Stukeley, writing of the Roman road from Sorbiodunum to Durnovaria, in his "Itinerarium Curiosum," 2nd edition, 1776, p. 188, thus describes the site:

"Its high ridge is then inclofed within a pafture juft at Woodyates, "then becomes the common road for half a mile, but immediately paffes "forward upon a down, the road going off to the right. I continued "the Roman road for two or three mile, where it is rarely vifited: it "is very beautiful, fmooth on both fides, broad at top, the holes remaining "whence it was taken, with a ditch on each hand: it is made of gravel, "flint, or fuch ftuff as happened in the way, moft convenient and lafting. "There are vaft numbers of Celtic barrows upon thefe downs, juft of "fuch manner and fhapes as thofe of Salifbury plain: at the firft and "more confiderable group I came to, there was a moft convincing "evidence of the Roman road being made fince the barrows. . . . One "form of thefe barrows, for diftinction fake, I call Druids (for what "reafons I fhall not ftand here to difpute:) they are thus. A circle of "about 100 feet diameter, more or lefs, is inclofed with a ditch of a "moderate breadth and depth: on the outfide of this ditch is a pro- "portionate *vallum*: in the centre of this inclofure is a fmall tump, "where the remains of the perfon are buried, fometimes two, fometimes "three. Now fo it fell out, that the line of direction of the Roman "road neceffarily carried it over part of one of thefe *tumuli* and fome "of the materials of the road are dug out of it: this has two little tumps "in its centre."

Dr. Stukeley's estimate of Ackling Dyke is as true now as when he wrote. Mr. Thomas Codrington, writing of this road, says: "This is perhaps

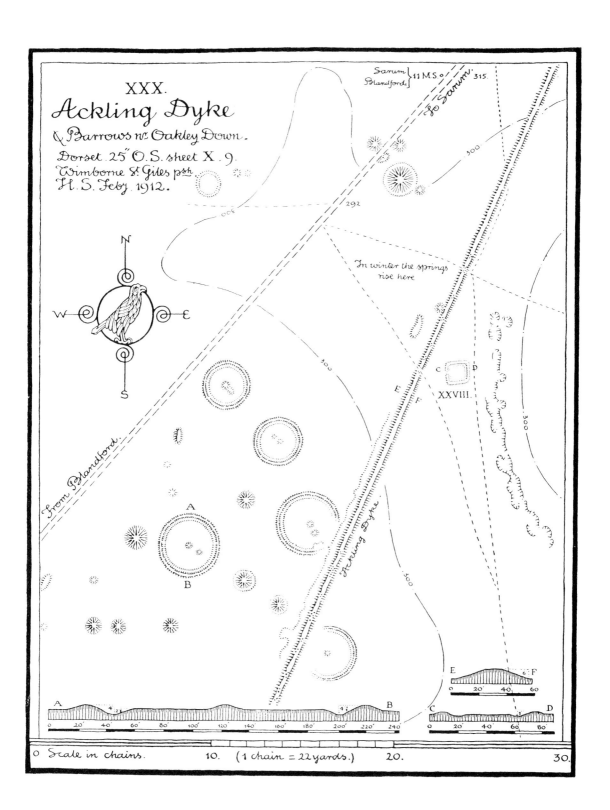

XXX.
Ackling Dyke
& Barrows nr. Oakley Down.
Dorset. 25" O.S. sheet X. 9.
Wimborne St Giles psh.
H. S. Feby. 1912.

Sarum } 11 M.S. o
Blandford }
to Sarum. 315.

292

In winter the springs
rise here

From Blandford.

Ackling Dyke.

XXVIII.

c d

E
F

A

B

E ...6... F

A ...3...   B

C   D

0 Scale in chains.   10.   (1 chain = 22 yards.)   20.   30.

" the most striking example of the embankment of a Roman road re-
" maining in the country. It runs for miles in a straight line in bold and
" sharp relief over the open down, and the magnitude of the work and
" its situation are alike imposing " (" Roman Roads in Britain ").

In " Excavations in Bokerly Dyke," vol. iii, p. 74, General Pitt-
Rivers gives two excavated sections of this Roman road at Woodyates,
about a mile to the North-East of the site we are describing : " The layers
" in the centre consisted of (1) surface mould, 5 inches ; (2) gravel with
" rounded pebbles, probably from patches of tertiary formation on
" Pentridge Hill, 6 inches ; (3) rammed chalk rubble, 6 inches ; (4) tertiary
" gravel again, 10 inches ; (5) rammed chalk, 6 inches ; (6) a single layer of
" nodular flints, laying on the old surface line. The total height from the
" old surface line to the top of the road was 3 feet."

The disk barrows shown in the plan are most beautiful in their
precise and circular forms. They are specially referred to by Mr. Charles
Warne, in " Celtic Tumuli of Dorset," as among the finest to be seen in
the county. Dr. Stukeley claimed disk barrows as Druid monuments, and
indeed called them Druid barrows, although he gave no evidence for his
assumption.

Dr. Thurnam, in his exhaustive account of " British Barrows,
especially those of Wiltshire and the adjoining Counties " (" Archaeologia," vol. 43), thus analyzes the results of excavations in disk barrows :
" As compared with the bell-shaped (barrows) they are probably of more
" recent invention, being more uniformly connected with the rite of
" cremation. The interments, with scarcely an exception, consist of the
" burnt remains, deposited in small dished graves scooped out in the chalk
" rock, and hardly ever inclosed in urns. Of the thirty-six disc-shaped
" barrows, of the exploration of which these are details, thirty-five
" contained interment after cremation. In thirteen of this number there
" were beads and other ornaments of amber, glass, and jet, often in such
" profusion as to justify the conclusion of Sir Richard Hoare that they
" are the burial-place of women ; especially as such objects are rare in
" the bowl-shaped barrows, and still more so in those of the bell-shape."

There would seem to be some ritual meaning in the unbroken circles
of bank and ditch surrounding these low mounds ; it may be that Dr.
Stukeley's name was not entirely wrong after all ; and these precise earthworks may express Druidical belief in the circular turn. In Borlase's
" Antiquities of Cornwall " will be found a curious chapter on the ancient
ritual of circular turns.

There is a small square earthwork, inclosed by a low bank with its
ditch on the inside, containing about a quarter of an acre, that lies in the

valley a few yards to the East of Ackling Dyke. In size it compares with the small camp on Handley Hill, excavated by General Pitt-Rivers in 1893, and proved to be of the Bronze Age, or of the early Roman period, and with the enclosure on Knighton Hill. But its situation and construction are quite different from either of these earthworks. They are on hill sides, this is in a valley, and their ditches are on the outside. In early Roman times we know that the chalk springs constantly rose above their present outflow on Cranborne Chase (*cf.* General Pitt-Rivers, " Excavations in Cranborne Chase," Woodcuts, vol. ii). Now, at mid-winter, both the ditch and the area of this valley enclosure are sometimes flooded. Accordingly, we can hardly suppose that it was made at a period when such conditions were constant throughout the year, so it seems possible that this enclosure is not of great antiquity.

Beyond this valley, to the East, towards Pentridge, the rising down is ridged with cultivation banks that seem to be ancient. Similar ridges may be seen on the sides of Bottlebush Down that face towards the North, suggesting more cultivation in the past than we now behold on these hills.

On the top of the rising down before mentioned runs a wide-banked track that Sir R. C. Hoare called " the Cursus," leading to the British settlement on Gussage Down. Here may be seen what appears to be a long barrow, actually part of the line of the North-Western bank of the " Cursus," suggesting that it was there, and utilized as a bank when the track was embanked by its makers.

CASTLE HILL, CRANBORNE

This earthwork is situated on a little hill that rises 300 feet above the sea North of the road from Cranborne to Edmondsham. It appears to belong to the Norman type of entrenchment known as the Motte-and-Bailey Castle—the Motte being the hillock or mound surrounded by the deep ditch, which would have been surmounted by a wooden structure, the fortress home of the feudal Baron; the Bailey, being the entrenched courtyard for the accommodation of his retainers. " Early Norman Castles of the British Isles," by Mrs. Ella S. Armitage, with numerous plans by Mr. D. S. Montgomerie, should be referred to for a clear statement of the origin of these castles, and of their essential difference of purpose from that of the Saxon " burhs " or boroughs, and of the British camps. " The Norman landholder " desired a safe residence for himself amidst a hostile peasantry," while "all the fortifications which

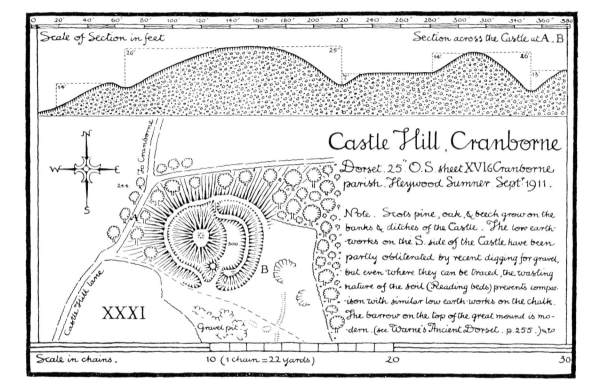

Scale of Section in feet

0 20′ 40′ 60′ 80′ 100′ 120′ 140′ 160′ 180′ 200′ 220′ 240′ 260′ 280′ 300′ 320′ 340′ 360′ 380′

Section across the Castle at A . B

26″   25″   14″   16″

14″   13″

Castle Hill, Cranborne

Dorset. 25″ O.S. sheet XVI 6 Cranborne parish. Heywood Sumner Sept.ʳ 1911.

Note. Scots pine, oak, & beech grow on the banks & ditches of the Castle. The low earth-works on the S. side of the Castle have been partly obliterated by recent digging for gravel, but even where they can be traced, the wasting nature of the soil (Reading beds) prevents compar-ison with similar low earth-works on the chalk. The barrow on the top of the great mound is mo-dern. (see Warne's Ancient Dorset. p. 255.) ⬦

To Cranborne

244.

Castle Hill Lane

XXXI

Gravel pit

500

B

Scale in chains.   10 (1 chain = 22 yards)   20   30

" we know to have been built by the Anglo-Saxons were the fortifications
" of society and not of the individual."

The following description given by Mrs. Armitage of this type of
earthwork applies to the entrenchments that we can now see on Castle
Hill, Cranborne: " They consist, when fully preserved, of an artificial
" hillock, 20, 30, 40, or in some rare instances 100 feet high. The
" hillock carried a breast-work of earth round the top, which in many
" cases is preserved; this breast-work enclosed a small court, sometimes
" only 30 feet in diameter, in rare cases as large as half an acre; it
" must have been crowned by a stockade of timber, and the representa-
" tions in the Bayeux Tapestry would lead us to think that it always
" enclosed a wooden tower. . . . The base of the hillock is surrounded
" by a ditch. Below the hillock is a court, much larger than the small
" space enclosed on the top of the mount. It also has been surrounded
" by a ditch, which joins the ditch of the mount, and thus encloses the
" whole fortification. The court is defended by earthen banks, both on
" the scarp and counterscarp of the ditch, and these banks of course had
" also their timber stockades, the remains of which have sometimes been
" found on excavation." . . . "It is clear that the man who threw up
" such earthworks was not only suspicious of his neighbours, but was
" even suspicious of his own garrison. For the hillock in the great majority
" of cases is so constructed as to be capable of complete isolation, and
" capable of defending itself, if necessary, against its own court."

These extracts give a vivid interpretation of the earthworks which
we are surveying. Access from the lower court or bailey to the mount
or motte can only have been obtained by means of a wooden bridge, or
ladder, and such means of access is graphically recorded in the Bayeux
Tapestry. These forgotten mounds and ditches yield a grim record of
the feudal system, when the great Baron's hand was against every man,
and when he even defended himself in his isolated motte against his
own retainers. On either side of the entrance to the lower court or
bailey at Castle Hill, Cranborne, there is a mound that was probably
surmounted by a wooden gate-house. Dr. Wake Smart's views as to this
earthwork are recorded in the " Dorset Field Club Proceedings," vol. xi.
He denies its (then) supposed British origin, but claims it as a Saxon
castle. Mrs. Armitage's researches show that such castles cannot be
assigned to the Saxon period.

The disconnected earthworks on the Edmondsham side of Castle
Hill may belong to an earlier British occupation of this site, but they are
so much wasted, owing to the soil (Reading Beds), and to gravel digging,
that conjecture is useless.

THE MIZMAZE ON BREAMORE DOWN

In Hereford Cathedral there is a thirteenth-century map of the world on which, amongst many strange geographical shapes, is figured the island of Crete, with a labyrinth plan, and the inscription " Laborintus " id est domus Dealli." The Breamore Mizmaze is similar to this Hereford labyrinth plan, and also to the following grass-cut mazes in England, namely, at Alkborough, Lincolnshire, at Broughton Green, Northants, at Ripon Common (now destroyed), and to the pavement mazes in French churches at S. Quentin and at Chartres, and to one that is incised on the Cathedral of Lucca.

The device of a labyrinth first appears in the fifth or sixth century B.C. on Cretan coins; and the earliest that we read of was " The Cretan " Labyrinth built by Dædalus, in imitation of a more ancient Labyrinth " in Egypt, by command of King Minos. First, it served as a prison for " the monster Minotaur, then as an architectural web to enclose Dædalus, " whence he was enabled to escape by aid of artificial wings " (Trollope, " Journal of Archaeology," vol. xv, from whom my further quotations also are taken). Herodotus describes an Egyptian architectural labyrinth at Lake Maeris. Pliny alludes to an architectural labyrinth for royal sepulchres both in the island of Lemnos and of Samos. Thus the early uses of labyrinths seem to have been as prisons and as burial-places.

The labyrinth at Lucca, already referred to, marks a new departure in the signification of the design. It is incised on the porch pier of the Cathedral. The centre is filled by Theseus and the Minotaur (nearly effaced), and at the side is the following inscription:

Hic quem Creticus edit Dedalus est Laberintus,
De quo nullus vadere quivit qui fuit intus,
Ni Theseus gratis Ariane stamine Jutus.

The Church now further appropriated the Pagan labyrinth, and instead of the Minotaur, " Sancta Ecclesia," or the Cross, was inscribed as the centre of the design, which was " deemed to be indicative of the " complicated folds of sin by which man is surrounded, and how impos- " sible it would be to extricate himself from them except through the " assisting hand of Providence."

Labyrinths were then frequently laid down in coloured marbles on church floors, e.g., at S. Maria Trastevere, Rome, at Ravenna, at S. Quentin, at Chartres, at Rheims (1240, destroyed 1794), at Amiens (1288, destroyed 1825), etc.

The Mizmaze on Breamore Down XXXII.

THE MIZMAZE ON BREAMORE DOWN

Later on these labyrinths were used as instruments of penance for non-fulfilment of vows of pilgrimage to the Holy Land, and were called "Chemins de Jérusalem." The pilgrims followed the windings of the Maze on their knees, and the centre was called "Le Ciel." Some of these labyrinths were destroyed because "children by noisily "tracking out their tortuous paths, occasioned disturbance during divine "service."

We have no ancient example of an ecclesiastical labyrinth in any English church, but our turf-cut mazes are undoubtedly of mediaeval ecclesiastical origin. They are situated near a church or a monastic settlement, and still remain at Breamore, Alkborough, Wing, Boughton Green, Saffron Walden, Sneinton, and St. Catherine's Hill, Chilcombe, etc.

The priory of St. Michael's, Breamore, was founded for Austin Canons about 1129, and the Mizmaze may be connected with the priory.

In Elizabethan times these English mazes would seem to have been secularized. Titania thus bears witness:

> The nine-men's-morris is filled up with mud;
> And the quaint mazes in the wanton green,
> For lack of tread, are undistinguishable.

The local name for them was "Julian Bower," "Troy Town" (Troi in Welsh means "to turn"), "Shepherd's Race," or "Mizmaze." The Renaissance garden mazes, in which the paths were misleading, have no relation to the labyrinth type, in which the paths always lead continuously to the centre of the design (cf. "Architecture, Mysticism and Myth," by Professor W. R. Lethaby).

DYKES AND DITCHES

BOKERLY DYKE

"TO everything there is a season, and a time to every purpose "under the heaven." So there is a time to see places—a time when they look their best—and the time to see Bokerly Dyke is when the sun is low in the West. The wayfarer from Salisbury to Blandford will then see the great scarp of the rampart defined by shadow; he will be able to trace at a glance the zigzag course of the Dyke over Martin Down and up Blagdon Hill, and to realize its size and continuity; thus he will see Bokerly Dyke rising and falling across the folds of the downs; a huge barrier casting its shadow in a lonely land—and he will wonder. Indeed it is a wonderful earthwork, and one that suggests desperate men and stern purpose to account for its construction.

Who made it? Why was it made? These are questions that we cannot fail to ask.

The excavations in Bokerly Dyke made by General Pitt-Rivers during the years 1888-90, and recorded in his third volume, give the best answer that can be found; but before I quote from this book, it may be interesting to note the varying opinions of previous antiquaries—who judged by superficial appearance, without the aid of excavation.

Dr. Stukeley, in his "Itinerarium Curiosum," 1st edition 1724, 2nd edition 1776 (p. 187), writing of the Roman road from Sorbiodunum to Durnovaria, thus refers to Bokerly Dyke (he calls it "Venndike"): "When it [the Roman road] has paſſed through the woods of Cranburn "Chaſe, and approaches Woodyates, you ſee a great dike and vallum "(Venndike) upon the edges of the hills to the left by Pentridge, to "which I ſuppoſe it gave name: this croſſes the Roman road, and then "paſſes on the other ſide, upon the diviſion between the hundred. The "large vallum here is ſouthward, and it runs upon the northern brink of "the hills; whence I conjecture it a diviſion or fence thrown up by the "Belgae before Caesar's time. . . . I pleaſed myſelf with the hopes of "obſerving the Roman road running over it, as doubtleſs it did originally;

" but juft at that inftant both enter a lane, where everything is dis-
" figured with the wearing away and reparations that have been made
" ever fince."

Dr. Guest, " Origines Celticae," 1850, " The Belgic Ditches," vol. ii,
p. 209: " As Badbury commands the valley where lay Vindogladia—
" which existing remains, as well as Itineraries, point out as the capital of
" the diftrict—and as Bokerly ditch was obviously intended as the northern
" boundary of this valley, it seems difficult to escape the conclusion that
" both Combe-bank (to the west of the Stour) and Bokerly-ditch were
" conftructed as parts of one design, and by the same people, and at the
" same, or nearly the same period. That people we may conjecture to be
" the Belgæ, and the period five or four, or, it may be only three centuries
" before the Christian æra."

Mr. Charles Warne, "Ancient Dorset," 1872, pp. 5 and 316: " I have
" no hesitation, however, in recording my deliberate opinion, that this
" mighty rampart owes its rise to the alarm produced by Cæsar's invasion
" of Britain." . . . " Bockley Dyke, in the N.E. of Dorset, may, indeed,
" from its position, be supposed to have been originally what in fact it
" now is—a boundary dyke: it may (and we believe it to) have been,
" however, a military work, conftructed as a barrier against impending
" invasion, and raised at a much later æra than when the Belgæ were
" dominant in the neighbourhood."

The Rev. William Barnes, "Dorset Field Club Transactions," vol. v,
1883, considered that " Bockley Dyke was a boundary." In vol. vi of
the same Transactions, 1884, Dr. Wake Smart agreed with Mr. Barnes
in considering "that Bockley Dyke was not a defensive work."

Such had been the contradictory opinions of antiquaries when
General Pitt-Rivers began his excavations at the Woodyates end of
Bokerly Dyke in the year 1888.

The first section cut by General Pitt-Rivers through Bokerly Dyke
was made South of the high road in the bend where the Dyke runs
nearly parallel to the road, and the evidence of the coins found in the
rampart and silting of the ditch, " extending from Gallienus to Honorius
" A.D. 395-423 proved that the dyke must have been made at the time
" of, or subsequently to, the departure of the Romans from the British
" isles A.D. 407. . . . It was evident that a Settlement must have existed
" on the ground before the Dyke was thrown up. The greater part of
" the coins were found in the lowest part of the rampart, in dark mould
" just over the old surface line, and it appeared quite certain that this
" mould must have come from the upper part of the ditch when the
" diggers threw up that part first, before they reached the chalk beneath,

" all of which was found overlying the mould in the rampart,. and con-
" taining comparatively few coins."

Subsequent excavation in the ground adjoining either side of the
Salisbury and Blandford road resulted in the discovery of a large Romano-
British settlement: and General Pitt-Rivers considered that here was the
site of the Roman station Vindogladia, in opposition to Sir R. C. Hoare's
opinion that this site was on Gussage Down. The latter settlement has
never been excavated, but has been much effaced by cultivation. All the
earthworks now visible on Gussage Down suggest pre-Roman work.
The balance of evidence at present obtained seems to be in favour of
Woodyates as the site of Vindogladia.

General Pitt-Rivers accordingly has proved Bokerly Dyke to be not
earlier than the end of the Roman occupation, by the evidence of his
excavations in one part of its course, that is fairly typical of its construction throughout. This was the period of the oncoming West Saxon. A
period of unusual emergency (and the Dyke is unusual)—Wansdyke
compares with it—this also General Pitt-Rivers proved by excavation to
be post-Roman; and the dykes on Charlton Down and White Sheet Hill
compare with it. Neither of these have been excavated, but their superficial measurements are suggestive. Bokerly Dyke, as interpreted by
General Pitt-Rivers' excavations, gives a standard of construction that will
help us in considering various other somewhat similar entrenchments
such as these. Opinion founded on superficial survey must always be
tentative and provisional, still reasonable surmise may encourage some
unknown seeker to excavate—to verify or to disprove—and taking
Bokerly Dyke as the standard of the sort of earthwork which the
Romano-British would have thrown up as a defence against the West
Saxon, we may conjecture that the following earthworks on Cranborne
Chase belonged to the same period, and had the same purpose. The
inside South-Eastern bank of Whitsbury Castle Ditches, the entrench-
ments of Clearbury Ring, Great Ditches (possibly), Charlton Down,
Hatt's Barn, Melbury Down, the inside bank of Hod Hill Camp, the
South-Eastern outer defences of Hambledon Hill Camp, and Half Mile
Ditch on White Sheet Hill.

All earthworks express passing requirements of life. Cranborne
Chase is seamed with low, rambling banks and ditches that dimly express
the detached requirements of British settlement life. Bokerly Dyke
expresses combination and a united resistance over a large area to a
common foe. The Hill-forts express isolated tribal defence against raids.
The continuous entrenchment of Bokerly expresses the unity of the
Romano-British inhabitants of Cranborne Chase in resisting the oncoming

GRIM'S DITCH

West Saxon. The unrecorded failure of this great Dyke is only testified by the abandoned sites of the British settlements, and by the existing sites of the valley villages, which mark the different choice of the West Saxon conquerors. Bokerly Dyke was a gigantic failure. The earthwork remains, but its purpose failed, and has faded into oblivion.

There is a branch of Bokerly Dyke—so-called in the Ordnance Survey—to which reference is made in my subsequent notes on Grim's Ditch. This may be traced running parallel to Bokerly Dyke, on the South-Western side, from the low earthworks near Bokerly Gap, up to the top of Blagdon Hill. Here, it turns eastward, and is apparently cut by Bokerly Dyke. Then it reappears, clearly defined, and similar to Grim's Ditch. With the exception of one gap (across the Tidpit Valley), its course may be traced to Damerham Knoll, and along its Eastern side, past the camp, on the far Eastern side of which it is lost in cultivation. I do not think that this branch has anything to do with Bokerly Dyke, but that it is a continuation of Grim's Ditch, and the section that I cut through this earthwork on Knoll Down confirmed this opinion. The missing link of its continuation from Damerham Knoll to Whitsbury has probably been destroyed by woodland grubbing, and by subsequent cultivation.

The small ditch and banks shown in the plan, running from Bokerly Dyke, Northward, to Vernditch Chase, was cut by General Pitt-Rivers near Martin Down Camp. He excavated 300 feet of both ditch and bank, with the result from the evidence of the pottery finds (p. 190, vol. iv) that this small trench was proved to be of the Bronze Age. It appears to belong to the boundary type of entrenchment. Near Vernditch this ditch abuts on a fragmentary entrenchment (on the Eastern side of the Roman road and of Vernditch Chase) which suggests pastoral usage. Towards the North this enclosure has been destroyed by cultivation.

GRIM'S DITCH

GRIM'S DITCH is a continuous earthwork—a ditch between two banks—that meanders up hill and down dale across the Eastern portion of Cranborne Chase. The length of its course is about fourteen miles. Its usual width is from 50 to 60 feet from outside to outside of its banks. One or other of the banks is frequently the higher, apparently without defensive purpose, but owing to the lie of the land. Where it is most perfect the superficial bottom of the ditch is from 5 to 6 feet below the higher bank. Its course twists and turns in a manner that is now unintelligible. In some places it is overgrown with hedgerows and timber trees; here the ditch is silted up with the woodland litter of years. In others the plough has cut away the banks, and only the ditch remains; while again elsewhere both banks and ditch have been ploughed and spread so thoroughly that the earthwork is quite obliterated, and its direction can only be traced by chalky bands—representing the spread banks—on either side of a single band of better soil—representing the spread silt of the ditch.

It will be seen on referring to the plan, that the course of Grim's Ditch from its beginning near Middle Chase Farm up to Vernditch Chase is now hardly discernible. This is owing to the grubbing, in addition to the ploughing, of this part of its course, which was formerly covered by woods, and also to the deeper cultivation here needed by the soil.

On the South-West of Vernditch Chase the double banks of Grim's Ditch appear to be equal in height. Elsewhere they vary, as already mentioned. Throughout the course of this far-reaching earthwork its construction does not suggest purposes of defence, of connection between British settlements, or of desire for concealment. It stops abruptly near Down Barn, a mile from Clearbury Ring, where it is cut at right angles by the track of a similar ditch that runs with a straighter course from Clearbury Down to Whitsbury Castle Ditches. There is no further trace of the ditch continuing Eastward towards and across the Avon Valley; but on Standlynch Down, three miles distant on the Eastern side of the Avon, it reappears, and runs up and along the Western slope of the Down to the high road between Salisbury and Romsey. Here it is intersected by deep trodden trackways that suggest pack-horse traffic of much later date than the Ditch.[1] Farther on cultivation has destroyed all traces of this earthwork, though the name of Grimstead, belonging to the hamlet close by, seems a fitting termination for Grim's Ditch.

Let us now return to the Ditch track that runs with a straighter course from Clearbury Down to Whitsbury Castle Ditches. It is most clearly defined on Court Down and on Breamore Down, after which it can be faintly discerned leading up to Whitsbury Castle Ditches. Here it apparently ends. There is no trace of it beyond, crossing the Rockbourne Valley, or mounting the slopes of Damerham Knoll, but at the Northern end of Damerham Knoll, on Knoll Down, a similar ditch between double banks recommences, and may be traced skirting a British

---

[1] Compare these with the sunken trackway that cuts the defensive dyke on Charlton Down, beside the steep hill above Donhead Hollow; and with those beside the Winchester and Morestead road, which also intersect a ditch of the boundary type that issues from the eastern side of the camp on St. Catherine's Hill.

settlement on Tidpit Common Down and a Long Barrow on Blagdon Hill, where apparently it is cut by Bokerly Dyke, after which its course runs parallel to the dyke, a few yards to the South-West, until it dies away among some low banks near Bokerly Gap. This ditch is marked on the Ordnance Survey as a branch of Bokerly Dyke, but it seems to be more probable that it is a continuation of Grim's Ditch. A section cut through the banks and ditch on Knoll Down supports this opinion, to which I shall refer later on. It is only owing to the exigencies of space that this branch of Grim's Ditch is included in the Bokerly Dyke plan.

The following extracts will show that the purpose of Grim's Ditch has been variously interpreted by antiquaries.

Sir R. C. Hoare in "Ancient Wiltshire, South" (published in 1812) p. 244, says: "I have frequently had occasion to mention the numerous " banks and ditches with which the unenclosed district of our county " abounds, and I have ventured to hazard an opinion, that they were not " all formed for the same purpose, but that some were designed for " boundaries, and others only for lines of communication between British " villages. I am inclined to think, that both Grymsditch and Bokerley " were intended for the former purpose, for they vary materially in their " mode of construction from those I have lately described in the Everley " station (on Salisbury Plain), and have a decided *vallum* on one side; " whereas in those banks which I denominate covered ways, or lines of " communication, the most discerning eye cannot distinguish on which " side the *vallum* is the highest, so equally is the ground thrown up on " each side. In Grymsditch I found the *vallum* on the east and north " sides during its whole course;[1] . . . I am the more inclined to think " that Grymsditch was a boundary ditch, as during its whole course, I " could not discover any British village on its borders."

Dr. Guest, in his paper on "The Early English Settlements in South Britain," reprinted in "Origines Celticae," vol. ii, p. 149, considers that Grim's Ditch was a boundary, and that the name, etymologically considered, shows that the Anglo-Saxons accepted its boundary usage.

In his paper on "The Belgic Ditches," same vol., p. 200, Dr. Guest makes a distinction from superficial appearance between British roads and boundary lines. The former he considered were represented by a ditch with low mounds on either side of it, the latter by a ditch with a mound on one side only, but he does not support his proposition by evidence. Grim's Ditch would certainly appear to come under his British road distinction, yet (from descriptions) he considered that it was a boundary.

[1] In this, my survey of one hundred years later disagrees.

CRANBORNE CHASE

Mr. Charles Warne, in "Ancient Dorset" (1872), describes the course of Grim's Ditch in his first chapter, under the heading of "The Boundaries of the Durotriges," while farther on, p. 29, under the heading of "The Roads and Trackways of the Durotriges," after opposing Sir R. C. Hoare's "covered way" theory—somewhat gratuitously, in this instance, for, as we have shown, Sir R. C. Hoare did not advance this theory as an explanation of Grim's Ditch—he considers that this, and such similar earthworks "may have served a double purpose, as.'ways'" "and boundary lines, or divisions between the property of contiguous" "tribes—Grim's Ditch is a road of this kind."

Dr. Wake Smart, the author of "A Chronicle of Cranborne," considered Grim's Ditch to be a British trackway ("Proceedings of the Dorset Field Club," vol. vi, 1884).

These varying opinions of learned antiquaries testify to the problem that is presented by Grim's Ditch.

It is partly this uncertainty that makes the survey of ancient earthworks such a fascinating study. There is always the possibility that we may wrest the secret from these humps and hollows in the downs, and that we may be able to get a better understanding of remote manners of life by the careful investigation of wasted landmarks. But generally we have to be contented—or discontented—with probability. And here I would say a word in favour of the advocates of this or that antiquarian probability. Of course antiquaries have again and again been mistaken in their surmises, but the man who never made mistakes, never made anything, and we only arrive at conclusions in matters of antiquity through processes of doubtful probing and questioning. Such inquiries are stimulated by having some working theory, and theory is merely an attempt to fit the facts. Even General Pitt-Rivers, whose genius was rooted in proof, burst into unwonted surmise with reference to Grim's Ditch (see "Excavations in Bokerly Dyke," Appendix A, vol. iii). Accordingly, my survey of Grim's Ditch, after following its course throughout, and after having cut sections in two places, has led me to advocate this probability as to its purpose, namely, that it was a tribal boundary, and that it was originally constructed by the Britons, without intention of defence, or of communication as a trackway.

Grim's Ditch is the principal example of similar earthworks that may be found throughout this district. None are so continuous, but many exist that resemble it in dimensions and construction. I think that it may be taken as a standard from which the purpose of such similar ditches may be inferred, *i.e.*, primarily for boundaries. Just as Bokerly Dyke gives another standard from which we may infer of the purpose of similar dykes, *i.e.*, primarily for defence. To suppose that these wandering earthworks of the Grim's Ditch type were prehistoric ways, is to misunderstand their sections as originally constructed, and to misconceive the needs of tribal life. The various tribes would need boundaries always, but lines of communication only occasionally, and then the ridgeways would be used. Laborious road-making belonged to a much later stage of civilization. Probably traders and raiders fared along the ridgeways from time immemorial, but there is no evidence of road-making in the pre-Roman period. We get a truer conception of British life on Cranborne Chase if we regard the various tribal settlements as isolated units, intermittently connected with the outer world by trade routes—either continuous or inter-tribal—otherwise self-contained, bent on their own livelihood, and determined to resist intrusion. We may wonder at the toil needed to make such a boundary as Grim's Ditch, but may we not claim equal wonder for the toil needed to make our hedgerow divisions, and banks and ditches? If we, individualists, need fences to mark our boundaries, so tribes would need still more emphatic boundary lines to mark their limits. Moreover, a much greater barrier would be needed than now to stay the tribal cattle, for their bones testify that they must have been more active than our cattle, their metacarpals and metatarsals being deeply grooved—like deer bones—to take the tendons that in our domestic cattle have gradually diminished owing to muscular disuse.

It appeared to me, in view of the conflicting opinions previously cited, that excavation would give the only evidence that could explain the usage of Grim's Ditch. If it had ever been used as a trackway, the floor of the ditch would assuredly be wide, and would show signs of trampling. Accordingly I cut a section across the ditch on Breamore Down, near Gallows Hill (in August 1911), by kind permission of Sir Edward Hulse, and another on Knoll Down, beneath Damerham Knoll (in October 1911), by kind permission of Sir Eyre Coote.

The evidence on Gallows Hill amounted to this, the bottom of the ditch was narrow—about 1 foot 6 inches across, smooth, and water-worn—owing to the fall of the ground. The sides were abrupt and unbroken. No relics of any sort were found, nor indeed expected, as there are no signs of primitive occupation here. On Knoll Down the evidence was stronger. The section of the ditch was similar to that on Gallows Hill (see plan), the bottom of the ditch was not worn by any user, and measured 1 foot 6 inches across. Five small sherds of probably pre-Roman pottery were found 3 feet down in the silt of the ditch, and three sherds of similar pottery were found on the down level under the South-Eastern bank. The mole castings about here had shown that such potsherds were below ground, and it seems that this earthwork was thrown up in pre-Roman times, such potsherds having littered the down surface when the bank was originally raised, and now having been discovered as stated. With the evidence of these sections, I think it impossible to suppose that this ditch can ever have been used as a trackway. But positive evidence of the period of its construction is still needed.

In the plan showing the section as excavated on Knoll Down, a layer of flints is represented lying above the silt and rubble that filled the bottom of the ditch. They were large and evidently collected— perhaps for the purpose of facing the steep chalk sides of the ditch banks, and thus reducing the wasting effects of the elements.[1]

Grim's Ditch is locally known as the Devil's Ditch. Forty years ago Mr. Charles Warne could obtain no direction from a countryman in response to his inquiry as to the whereabouts of Grim's Ditch, but information was at once forthcoming when he had assented to—" you mean Devil's Ditch." I have had precisely the same experience.

ENTRENCHMENTS ON WHITE SHEET HILL RIDGEWAY, AND ON THE OXDROVE RIDGEWAY

THE map of Cranborne Chase shows two high chalk ridges running East and West, parallel to each other, with the Ebble stream flowing Eastward down the intervening valley. The Northern ridge stretches from Harnham Hill to White Sheet Hill. The Southern ridge is named after the oxdrove which runs as a wide grass drove-way from Knighton Wood to Charlton Down.

We naturally suppose these continuous upland ridges to have been the ways of ancient traffic. Probably they were, but proof is difficult, and there is room for doubt in the courses of these ways. Certainly the existing Oxdrove Ridgeway track crosses the banks of the large enclosure above Chickengrove Bottom, showing that its present course at this place is not original. While again, on the White Sheet Hill ridge, East of Crockerton Firs, the old road cuts through some earthworks belonging to a British settlement on Swallow Cliffe Down, again showing that this cannot be accepted as the actual way of the ancient traffic or trade route.

[1] The entrance to Fosbury Camp, on the Vernham's Dean side, is guarded by steep banks that appear to have been faced with rammed flints.

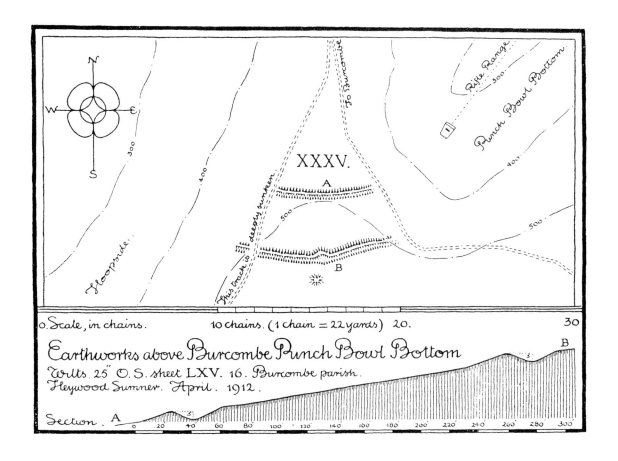

Earthworks above Burcombe Punch Bowl Bottom

Wilts. 25" O. S. sheet LXV. 16. Burcombe parish.
Heywood Sumner. April. 1912.

Still we may suppose that during prehistoric times, when the valleys were swamps, these ridges afforded the best opportunity for such wayfaring as was needed, and the Oxdrove Ridgeway track, as it passes above Ashcombe, as it winds round Win Green, and as it descends the sunken way beside the zigzag hill, seems to mark the ancient line of traffic.

The Ordnance Survey marks many banks and ditches crossing these two ridges from scarp to scarp. As shown on the map, their purpose seems to be obvious—they must have been made to stop an enemy coming along these ridgeways—and this enemy has been supposed to have been the West Saxon, whose advance Westward, from the Avon Valley, dates somewhere between A.D. 552, battle of Searo-byrig (Old Sarum), and A.D. 577, battle of Deorham, near Gloucester. But when these earthworks have been examined their evidence is not so simple. Let us take them in order from East to West, first from Harnham Hill to White Sheet Hill, and then on the Oxdrove Ridgeway to Charlton Down.

At the outset it may be well to state that in this chalk soil there are no modern ditches to confuse our survey, while modern fence banks invariably assert their origin by their narrow mound forms, compared with the broad, round-topped forms of ancient banks. The date of the following earthworks is dubious, but their claim to antiquity is sure.

(1) Two lines of bank and ditch cross the down spur that runs Northward above Burcombe (see plan). A deeply sunk trackway cuts them both on the Hoopside scarp of the down. The earthworks appear to guard this way—as on Buxbury Hill (7 *infra*)—against an enemy coming down the ridge towards the valley, the bank being on the North side. There is an unusual kink in the centre of the upper bank that makes a semicircular widening in the even run of the ditch, the purpose of which is debatable. This is not shown on the Ordnance Survey Map.

(2) About a quarter of a mile South of the above, beside three barrows on the edge of Hoopside, a ditch begins, with wasted banks of the boundary type. It runs Eastward, up the down, for about a hundred yards, and then dies away under cultivation. Sir R. C. Hoare, in " Ancient Wiltshire, South," shows the continuation of this ditch across the upland.

(3) A ditch between two banks, of which the Western is the larger, begins on the Northern down scarp near Compton Hut, but it is soon lost under cultivation, and cannot now be traced across the ridge.

(4) About half a mile East of Chiselbury Camp a ditch between low banks may be traced for a few yards, both on the Northern and Southern scarps of the ridge, but cultivation has effaced all trace of its course between these two fragments.

(5) See Chiselbury Camp, p. 26.

(6) A well-defined ditch between banks of equal height, named Row Ditch (see plan), crosses the ridge on Sutton Down. It is unusually straight in its course, and precise in its construction.

(7) Buxbury Hill projects from the Northern side of the ridge into the valley of the Nadder, and, as this down spur sinks, it is crossed by a bank and ditch which compare with those above Burcombe (see plan). A sunken trackway (named Buxbury Hollow) worn far below the level of the ditch, leads down this hill. Converging down-tracks unite at the Southern intersection of this sunken way and the entrenchment. This earthwork suggests defence against an enemy advancing towards Castle Ditches Camp—a mile distant to the North—and the sunken way suggests either the constant trampling of cattle between the downs and the valley, or pack-horse traffic of later times.

(8) Just beyond the site of the British settlement on Swallow Cliffe Down a rambling ditch between two banks (of which the Eastern is the higher) crosses the ridge. It seems to be a defence connected with the British settlement, and is different in construction and course from Row Ditch (see plan).

(9) Near Crockerton Firs, on the down spur leading to Alvediston, a ditch between two banks, of which the Northern is the higher, crosses the down (see plan). It seems to be a slight defence against an enemy coming up from the Ebble Valley.

(10) About half a mile West of Crockerton Firs a ditch, between two banks of much greater height than any of the preceding (see plan), crosses White Street Hill. It is locally known as " Half Mile Ditch." The larger bank is on the Western side. It compares with the earthworks that cross Charlton Down, Hatt's Barn, Fontmell Down, Melbury Hill, and, in a lesser degree, with Bokerly Dyke. It seems probable that this unusually formidable defence was thrown up by the Romano-British as a barrier to stop the oncoming West Saxon. This probability, however, suggested by superficial appearance and comparison with Bokerly Dyke, can only be tested by the spade.

(11) A ditch between two low banks runs East and West across a Southern spur of White Sheet Hill—from the top of Berwick Combe to the down scarp above Ferne. The Southern bank is the higher of the two, and the ground falls from North to South. This may have been a boundary ditch, it does not suggest defence. Here ends White Sheet Hill.

Now let us cross over the Ebble Valley and survey the banks and ditches on the Oxdrove Ridgeway, taking them in order from East to West.

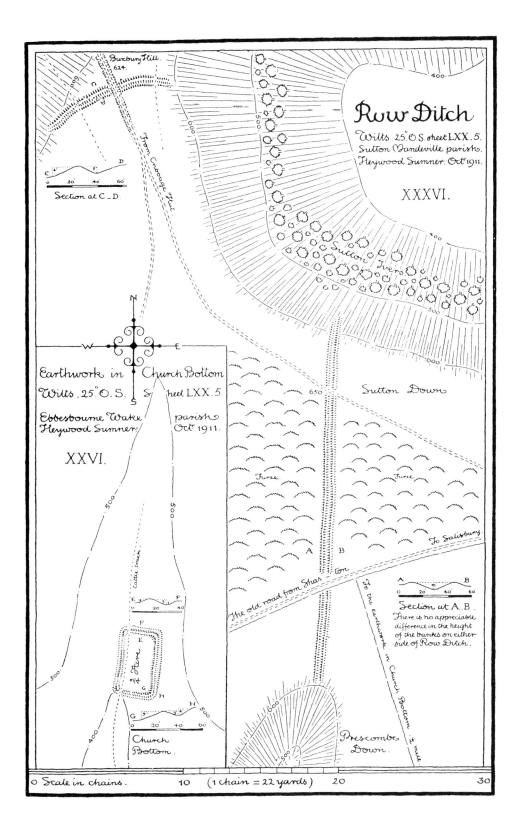

Row Ditch

Wilts. 25" O.S. sheet LXX. 5.
Sutton Mandeville parish.
Heywood Sumner. Oct 1911.

XXXVI.

Section at C_D

Earthwork in Church Bottom
Wilts. 25" O.S. Sheet LXX. 5.
Ebbesbourne Wake parish
Heywood Sumner. Oct 1911.

XXVI.

Section at A.B.
There is no appreciable
difference in the height
of the banks on either
side of Row Ditch.

Church Bottom.

Prescombe Down.

Scale in chains.   10   (1 chain = 22 yards)   20   30

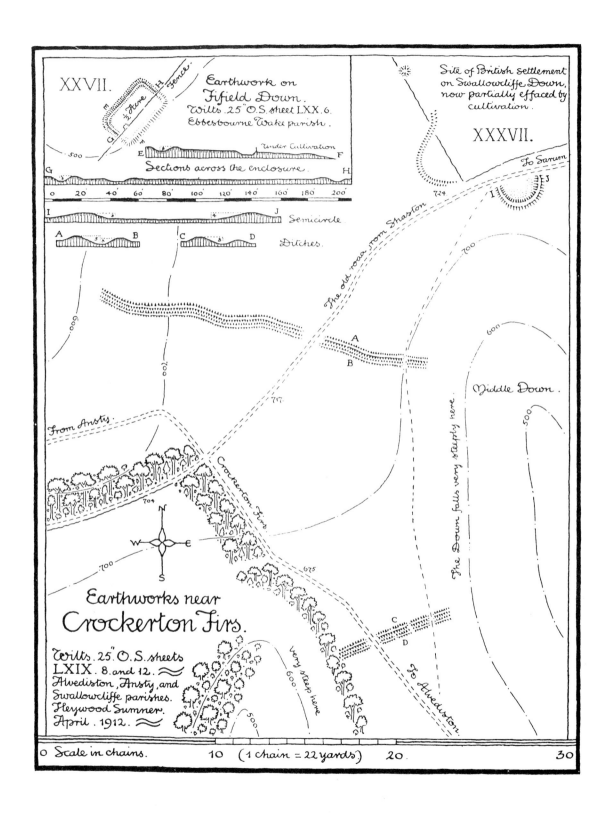

XXVII.

Earthwork on
Fifield Down.
Wilts. 25" O.S. sheet LXX. 6.
Ebbesbourne Wake parish.

½ Acre

fence

500

E · · · · · · · Under Cultivation · · · · · · · F
Sections across the enclosure.
G · H

0   20   40   60   80   100   120   140   160   180   200'

I · · · · · · · · · · · · · · · · · J   Semicircle

A · · · · B   C · · · · D   Ditches.

Site of British settlement
on Swallowcliffe Down,
now partially effaced by
cultivation.

XXXVII.

To Sarum

The old road from Shaston

724

700

600

A
B

717

Middle Down.

500

The Down falls very steeply here.

I   J

From Ansty

600

700

704

Crockerton Firs

N
W   E
S

700

675

C
D

To Alvediston.

Earthworks near
Crockerton Firs.

Wilts. 25" O.S. sheets
LXIX. 8. and 12.
Alvediston, Ansty, and
Swallowcliffe parishes.
Heywood Sumner.
April. 1912.

very steep here

600

500

0   Scale in chains.   10   (1 chain = 22 yards)   20.   30

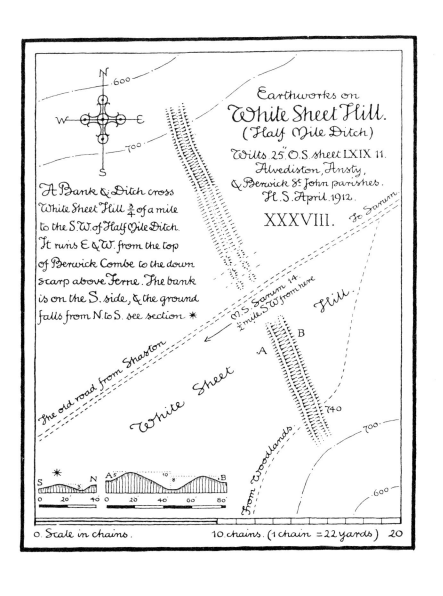

Earthworks on
White Sheet Hill.
(Half Mile Ditch)
Wilts. 25ʺ O.S. sheet LXIX 11.
Alvediston, Ansty,
& Berwick St John parishes.
H.S. April 1912.

XXXVIII.

A Bank & Ditch cross
White Sheet Hill ¾ of a mile
to the S.W. of Half Mile Ditch.
It runs E & W. from the top
of Berwick Combe to the down
scarp above Ferne. The bank
is on the S. side, & the ground
falls from N. to S. see section ✳

The old road from Shaston

M.S. Sarum 14.
½ mile S.W from here

To Sarum

White Sheet

Hill.

From Woodlands

.600

.700

740

700.

600.

A

B

S   N
0   20   40

A   10   B
8
0   20   40   60   80

0. Scale in chains.   10. chains. (1 chain = 22 yards)   20

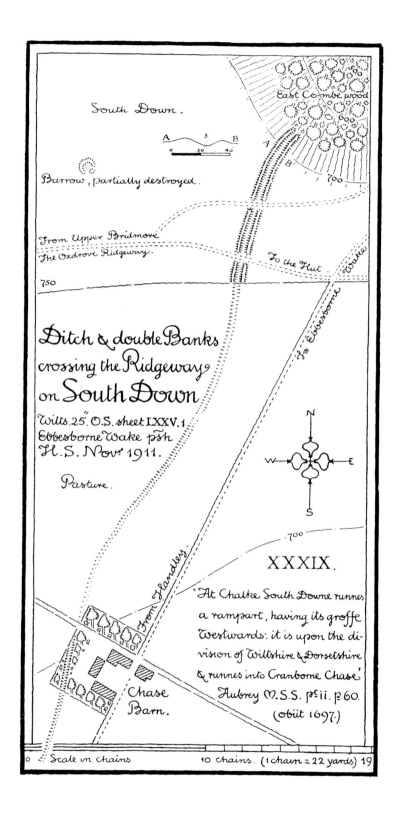

South Down.

A 5 B
0 10 40

Barrow, partially destroyed.

East Combe wood

600

700

From Upper Bridmore
The Oxdrove Ridgeway.

To the Flut

750

To Ebbesborne Wake

Ditch & double Banks
crossing the Ridgeway
on South Down

Wilts. 25". O.S. sheet LXXV. 1
Ebbesborne Wake p͞sh
H.S. Novr 1911.

Pasture.

N
W   E
S

.700

XXXIX.

'At Chalke South Downe runnes
a rampart, having its groffe
Westwards: it is upon the division of Wiltshire & Dorsetshire
& runnes into Cranborne Chase.'
Aubrey M.S.S. p͞t ii. p 60.
(obiit 1697.)

From Handley.

Chase
Barn.

0   Scale in chains   10 chains. (1 chain = 22 yards) 19

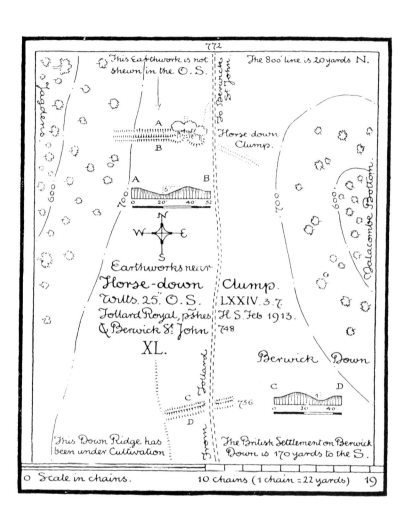

772

This Earthwork is not shewn in the O.S.

The 800' line is 20 yards N.

To Berwick St John

Horse-down Clump.

A
B

A · B
0   20'   40   52'

N
W · E
S

Earthworks near
Horse-down Clump.
Wilts. 25" O.S.   LXXIV. 3. 7.
Tollard Royal, pshes   H. S. Feb 1913.
& Berwick St John   748
XL.

Berwick   Down

C   D
0   20   40

756

from Tollard

C
D

This Down Ridge has been under Cultivation

The British Settlement on Berwick Down is 170 yards to the S.

o Scale in chains.   10 chains (1 chain = 22 yards)   19

ENTRENCHMENTS ON THE OXDROVE RIDGEWAY

(1) Great Ditch Banks, near East Chase Farm, have been ploughed over and spread by cultivation; so much so that it is now impossible to define their course; but their name, and the undulations that remain on the arable land suggest that this must have been a great earthwork. It may have been of the Bokerly Dyke type and period, but there is no superficial connection that can now be traced between these two ditches—Bokerly being about half a mile South of Great Ditch Banks. There are signs of this earthwork crossing the present Oxdrove Ridgeway; beyond that it has been effaced by cultivation.

(2) On South Down a well defined ditch between two banks of about the same height (see plan) begins on the Northern scarp of East Combe Wood.[1] It follows a rambling course, North-West to South-East, across a pasture where it has been partially effaced by cultivation, to Chase Barn. It is more clearly defined in the belt of trees to the West of Chase Barn, and dies away about fifty yards from this belt of trees in a pasture gently sloping to the South.

(3) A ditch between two banks begins to cross the Ridgeway on the North-Western outside of Stone Down Wood, but is almost immediately lost under cultivation.

(4) A ditch between two banks runs for a few yards up the verge of the down scarp above East Ivers Wood, and then disappears entirely under cultivation.

(5) About a mile farther on to the West a long ridge spur branches off to the South, above Ashcombe, leading down to Tollard Royal village. At Horsedown Clump a bank and ditch cross this ridge from the top of Jagdens to the top of Malacombe Bottom—a distance of about 150 yards. The bank is on the South side. This earthwork is not shown in the Ordnance Survey.

Two hundred and fifty yards farther to the South, on the top of Berwick Down, there is a fragment remaining of another bank and ditch which presumably also crossed the ridge, but which has mostly been destroyed by cultivation. From what remains of this earthwork we may infer that it was smaller than the one above. Its bank is on the North side. The construction of these two scarp to scarp entrenchments is unusual.

There are two other instances on Cranborne Chase of two parallel scarp to scarp earthworks with an intervening space between them—on Hoopside, above Burcombe, and on Fontmell Down—but in both these instances their banks face the same way—uphill. Here the bank at

---

[1] In Thomas Aldwell's "Mappe of Cranburne Chace," 1618, this wood is spelt "Iscombe," and it is still thus pronounced locally.

Horsedown Clump faces uphill, while the fragmentary bank on Berwick Down faces downhill. There is a large British settlement site on the South-Eastern side of Berwick Down.

(6) About half a mile to the West of Win Green, a ditch between two high banks (see plan) crosses Charlton Down from scarp to scarp, running North and South. The Western bank is the higher of the two. It is similar to Half Mile Ditch on White Sheet Hill, and its purpose was probably the same. The place-name of Win Green (half a mile to the North-East) is suggestive of the Anglo-Saxon—Win, A.S. *war*. The deeply sunken trackway that mounts the hill from Donhead passes through this earthwork, and the bottom of the trackway is much below the bottom of the ditch, *i.e.*, the trackway was made and worn subsequently to the making of this Charlton Down defensive barrier. Probably this is a pack-horse track. Similar cuttings of ancient ditches by pack-horse ways may be seen on Pepper-Box Hill above Alderbury, and beside the Morestead road, near Winchester.

(7) A mile farther on to the West, near Hatts Barn, a similar earthwork may be traced across the down ridge. In the pasture near Hatts Copse there are two slight banks on the Western side of the big ditch, instead of a single high bank. Elsewhere this earthwork is wasted by cultivation and tree planting, but from what remains it seems to belong to the Bokerly type. On the right of the road (going West) it is almost obliterated, and is ignored by the Ordnance Survey, but it can be traced to a little fir plantation above the deep combe that runs into the folds of Melbury Down. From here to Coney Hall Hill, above Melbury Village, the scarp of the down presents a natural defence, then, on the Western side of the old high road from Shaftesbury to Blandford, a ditch between wasted banks reappears, and runs along the North-East scarp of Compton Down, across the col that joins Compton Down to Melbury Hill, and here, half-way up the hill it cuts another bank and ditch at right angles— a defence that guards the South-Eastern approach of the summit from side to side. Then it dies away to a mere ledge on the precipitous slope of the punch-bowl re-entrant from Melbury Village, but farther on resumes its former section, and continues a somewhat rambling course, as before, till it dies away above the lower road from Shaftesbury to Blandford.

Throughout its course this entrenchment appears to have been thrown up against an enemy coming from the North-East, and (together with the natural defence of the steep down scarp, and with the Hatts Barn and Melbury Hill earthworks) defends a stretch of country for about three miles.

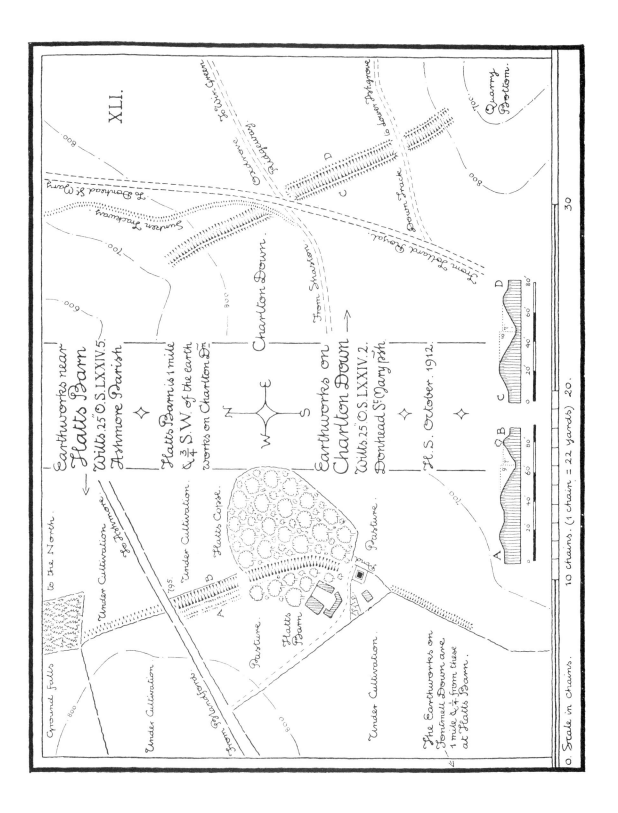

XLI.

Earthworks near
Flatts Barn
Wilts. 25" O.S. LXXIV. 5.
Ashmore Parish

Flatts Barn is 1 mile
& 3/4 S.W. of the earth
works on Charlton Dn.

Charlton Down.

Earthworks on
Charlton Down
Wilts. 25" O.S. LXXIV. 2.
Donhead St. Mary Pish.

H. S. October. 1912.

from Swanton

To Donhead St. Mary
Sunken Trackway

To Win Green
Ridgeway

to Lower Park grove

From Tollard Royal.

Down Track

Quarry
Bottom.

Ground falls
to the North.

Under Cultivation
to Ashmore

Under Cultivation.

Flatts Copse.

Under Cultivation.

795

Pasture.

Flatts
Farm

Pasture.

Under Cultivation.

The Earthworks on
Fontmell Down are
1 mile & 1/4 from those
at Flatts Barn.

from Shandsfords

A   B

C   D

Scale in chains.   10 chains. (1 chain = 22 yards) 20.   30

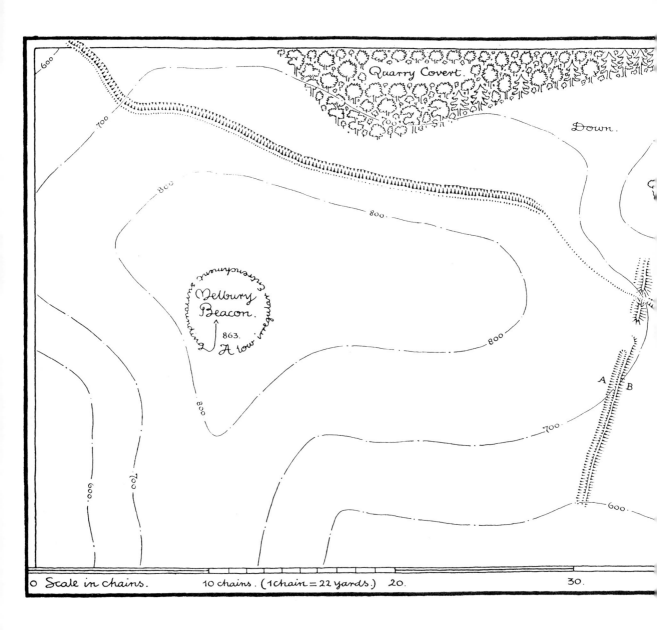

Quarry Covert

Down

Melbury
Beacon.
863.
A low irregular
entrenchment

A

B

600
700
800
800
800
800
700
600
700
600

O Scale in chains.   10 chains. (1 chain = 22 yards.)   20.   30.

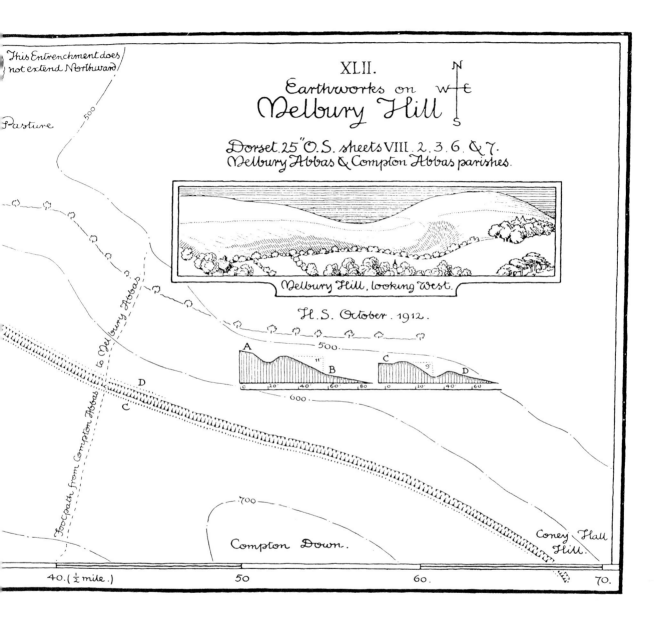

XLII.

Earthworks on

Melbury Hill

N W E S

Dorset. 25" O.S. sheets VIII. 2. 3. 6. & 7.
Melbury Abbas & Compton Abbas parishes.

Melbury Hill, looking West.

H. S. October. 1912.

This Entrenchment does
not extend Northward

Pasture

500

to Melbury Abbas

Footpath from Compton Abbas

A
B
500
11
0 20 40 60 80

C
9
D
10 20 40 60

D
C
600.

700.

Compton Down.

Coney Hall Hill.

40. (½ mile.)   50.   60.   70.

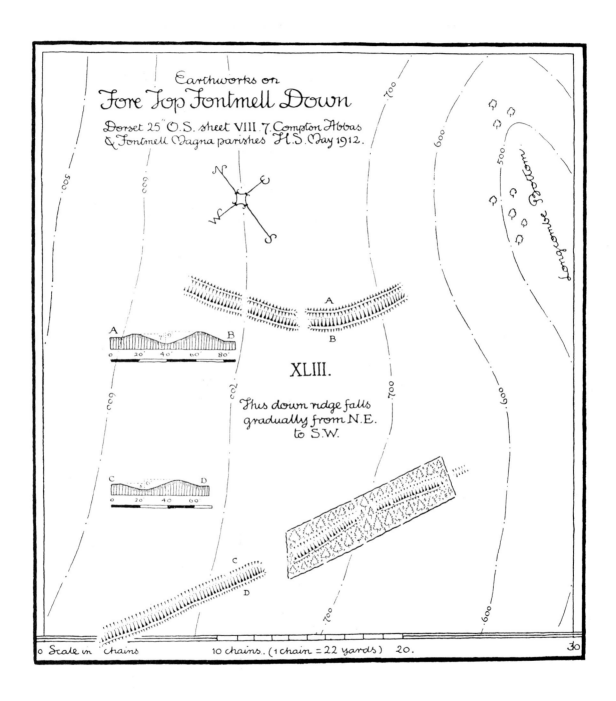

Earthworks on
Fore Top Fontmell Down

Dorset 25″ O.S. sheet VIII.7. Compton Abbas
& Fontmell Magna parishes. H.S. May 1912.

XLIII.

This down ridge falls
gradually from N.E.
to S.W.

0 Scale in chains   10 chains. (1 chain = 22 yards) 20.   30

(8) On Fore-top, Fontmell Down, two lines of banks and ditches (see plan) cross the down from scarp to scarp, separated from each other by an intervening space of level ridge 260 paces apart. There are gaps in the centres of both earthworks. In both, the higher bank is on the South-West side. The upper (Northern) line of defence is considerably stronger than the lower (Southern).

(9) A ditch between two banks begins on the verge of the scarp of Longcombe Bottom, and ends at the top of Fontmell Wood. On both sides the ditch is fairly discernible, but its banks have been partly ploughed away, and both banks and ditch have been quite effaced across the ridge. You may look in vain for any traces of this earthwork beside the old Shaftesbury and Blandford highway, which runs along the top of this cultivated hill-top. Yet it was one of the landmarks named in Thomas Aldwell's " Mappe " (1618), as limiting the inner bounds of the Chase at this point—Tennerley Ditch—and so we may suppose that Tennerley Ditch—a place-name now forgotten—was a notable earthwork when the inner bounds were delimited.

One fact emerges clearly from the foregoing survey, namely, that the big entrenchments on these two ridges are invariably raised against an enemy coming from the East, while the small banks and ditches seem to vary according to the lie of the land and to the requirements of British settlements. It seems probable that the former were raised by the Romanized Britons against the oncoming West Saxons, and that the latter were pre-Roman,[1] and used primarily for boundary purposes and cattle stops.

There is an entrenchment South of Ashmore (see map of Cranborne Chase) that seems to belong to the Grim's Ditch boundary type. It has been much spread by cultivation, but is fairly well preserved on the Eastern side of Mudoak Wood, where its Southern bank—which throughout appears to be the larger—rises 6 feet above the bottom of the ditch, and where its all-over measurement is 56 feet. The Roman road from Badbury Rings intersects this entrenchment, suggesting that the road was made at a later period than the entrenchment.

[1] See "Excavations in Cranborne Chase," vol. iv (Gen. Pitt-Rivers), where the results are given of the excavation of a ditch that seems to have been of the boundary type near Martin Down Camp.

BRITISH VILLAGE SITES

THE earthworks that denote the British village sites on Cranborne Chase do not specially appeal to superficial observation. Hill-top camps are always impressive. Barrows crown the ridges as landmarks. Dykes cast continuous shadows that call attention to their course as they rise and fall across the Downs. But the low, random humps and hollows that remain as the surface appearance of a British village site are confusing, intricate, and without apparent purpose. Indeed, so slight are their relief and depression that a plan really enables a better idea to be formed of their disposition and scope, than a wandering visit of inspection. I do not think that a sightseer would be much impressed by any of the British settlements on Cranborne Chase. There is so little to be seen. But if their rambling forms and their recurring features are grasped by a preliminary study of their plans, we may know what to look for, and then these slight earthworks will obtain a new meaning, and will indicate the manner of life and the habitations of our forefathers.

And how many of these settlements there are! Truly this area of Cranborne Chase was once a desired land. Woodcuts, Rotherley, Berwick Down, Fontmell Down (?), Compton Down (?), Woodyates, near Great Ditch Banks, Middle Chase Farm, Marleycombe and Chickengrove, Knighton Hill, Rockbourne Down (?), Damerham Knoll (?), Tidpit Common Down, Blackbush Down, Gussage Down, Oakley Down, Oakley Lane (or Minchington Down), Chettle Down, Tarrant Hinton Down, South Tarrant Hinton Down, Blandford Race-Down, Horse Down, near Crichel, and Swallowcliffe Down. A long list—yet how many more have been effaced by cultivation? Indeed, a census of Cranborne Chase in British times might be rather surprising to us nowadays.

These open British village sites seem to belong to a period when prehistoric life was lived under more settled conditions than obtained when the hill-top camps were made. Their humps and hollows suggest shelter from rough weather. Their inner, rambling ditches suggest careful

68

drainage. Their outer banks and ditches suggest cattle pens, wolf barriers, and boundaries, and perhaps, where they are duplicated and triplicated, defences against hostile raiders. The purpose of the low banks and ditches that radiate from the settlements on Gussage Down, Middle Chase Farm, South Tarrant Hinton Down, Tarrant Hinton Down, and Blandford Race-Down is debatable, but their superficial appearance suggests pastoral and boundary purposes. It may be conjectured that, when life became so far secure that it was safe to live in these slightly defended British settlements, the hill-top camps were only rarely needed for defence and refuge, and if the boundary banks and ditches led up to them, it would be because they would always be in use for purposes of folding cattle and sheep. Whether such conjecture is correct or not, the fact remains that the principal British settlements of Cranborne Chase did not shelter themselves close beside the neighbouring protection of the great hill-top camps. We may regard such open village settlements as a step forward in civilization. The changes wrought by time during historic periods are obvious and accepted, but we may need a reminder that prehistoric time was also subject to change. Life did not remain entrenched behind defensive banks and ditches for a period as long as our own era. Man does not live by stone-throwing, and livelihood, of the herdsman or of the agriculturist, gradually compelled changes. We have reason to assume, and the evidence of these British settlements shows, that there were periods in prehistoric times when the hill-top camps became old fashioned, and when pastoral and agricultural life was safely led in these open British village sites.

Three of these open British village sites on Cranborne Chase were thoroughly excavated by General Pitt-Rivers—Woodcuts (1884-5), Rotherley (1885-6), and Woodyates (1889-90). All three revealed a curious absence of definite plan. The recurring features common to all of them, and that we can recognize in the superficial appearance of the other British sites, were—drains, to get rid of surface water; pits, either for refuse or to get chalk for the land; mounds, that are not barrows; small circular or semicircular embanked depressions; and entrenched enclosures—presumably cattle pens or sheep folds—with other enclosures of a square or oblong type, adjoining the settlements, and bounded by scarps that are commonly called cultivation banks.

Pastoral care and agriculture are suggested as the pursuits of the makers of these earthworks. Many of the cattle enclosures must have been destroyed by cultivation, and in one instance, at Bussey Stool Park, I have seen such destruction in process, alas! But still, enough remain to show their frequency, and their connection with the life of these British settlements.

For a detailed description of this life as exemplified by the relics found during the excavation of these three sites, the reader is referred to vols. i, ii, and iii of Gen. Pitt-Rivers' " Excavations in Cranborne Chase," but the following extract from vol. iii will give a general idea of the civilization and manner of living of the Romanized Britons at Woodcuts and Rotherley. " Woodcuts, or rather a portion of it, was surrounded by an " entrenchment of slight relief . . . and at Rotherley, a portion of the " village was separated from the rest by a circular surrounding ditch, " similar to others which have been several times noticed in British " villages elsewhere, and which have been rather rashly assumed to be " sacred circles ; but no confirmation of this was produced by the excava- " tions—the circle, on the contrary, appeared to have been occupied in " the same manner as the rest of the village. In Woodcuts, three hypocausts " of T-shaped plan were found, which were probably British imitations " of Roman hypocausts for warming rooms by flues beneath the floors. . . . " The houses must have been built of dab-and-wattle, and, by means of " some of the fragments of plaster, which had been hardened by fire, and " upon which the impression of the twigs had been preserved, it was " possible to ascertain the exact thickness of the walls and the construc- " tion of the wattle work. Timber was also used in the construction of " the houses, as appears probable from the large number of iron nails, of " a size suitable for fastening beams of wood, and also from a number of " cramps of the kind now used for fastening timber together. Besides the " dab-and-wattle houses, which were probably round, some other houses " must have been made with flat sides, plastered and painted. This better " class of houses were peculiar to one quarter in Woodcuts, which, from " the quality of the other objects found in it, appears likely to have been " a rich quarter. The pits were probably used to contain refuse, and, " after being filled up to the top were subsequently used for the inter- " ment of the dead. The dead were not interred in these pits only, but " also in the drains, after they had been filled up to the top with earth, " a practice which, if not confined to this district, has, at any rate, not " been found elsewhere to such an extent as to lead to the inference that " it was a widely-spread British custom. It was a custom that is highly " favourable to anthropological research, as the skeletons are, by this " means, more clearly identified with the relics of the every-day life of " the inhabitants, than when they were interred in cemeteries, or tumuli, " at a distance from the places where they lived. . . . Their horses, oxen, " and sheep were of small size, the horse rarely exceeding the size of our

" Exmoor pony, viz. 11 hands 2½ inches. The oxen resembled our Kerry
" cow in size, but our Shorthorn in the form of its horns; and the sheep
" were of a long slender-legged breed, the like of which is only to be
" found at present in the Island of St. Kilda. The pig, as is always found
" to be the case in early breeds, that were but slightly removed from the
" wild boar, was of large size, with long legs and large tusks. The dog
" varied from the size of a mastiff, to that of a terrier. They ate the
" horse, and lived chiefly upon domesticated animals, but few remains of
" deer having been found in their refuse pits, from which, and from the
" absence of weapons generally, we may infer that they were not hunters,
" but that they lived a peaceful, agricultural life, surrounded by their
" flocks and herds. Their tools were iron axes, knives, and saws, only one
" or two small spear-heads having been found. They spun thread, and
" wove it on the spot, and sewed with iron needles. They grew wheat in
" small enclosures surrounding their villages, and ground it upon stone
" querns, and by measuring the number of grains to the cubic inch, it
" was found that their wheat was little, if at all, inferior to ours, grown
" at the same levels. They shod their horses with iron, and produced fire
" with iron strike-lights and flint. They cut their corn with small iron
" sickles, probably close to the ear, and it appears probable that they
" stored it in small barns, raised upon four posts, to preserve it from
" vermin. Their pottery was of various qualities, some harder and better
" baked than others: some vessels perforated as colanders, some in the
" form of saucers with small handles, some basin-shaped, others pitcher-
" shaped, others in the shape of jars, and vases of graceful form; and
" judging by the number of pots perforated with large holes on the
" bottoms, or sides, and having loops for suspension on the upper part,
" with large open mouths, it would appear probable that they made use
" of honey largely in their food, and that these vessels were employed for
" draining it into other vessels, from the honeycomb. . . . Judging by the
" slight trace of their habitations that remained, and the small size of
" them, and the apparently careless way in which they buried their dead,
" one might suppose that they lived in a poor way, and died unregretted
" by their friends; but, on the other hand, there were indications of
" comfort, and even of refinement. There were found fragments of red
" Samian ware of the finest quality, and highly ornamented, which, at
" that time, and in this country, was probably equivalent to our china;
" and a few fragments of pottery with green and yellow glaze, which
" was of extreme rarity amongst the Romans. They had chests of
" drawers, in which they kept their goods, which were decorated with
" bronze bosses, and ornamented with tastefully-designed handles of

71

" the same metal. They had vessels of glass, which implies a certain
" degree of luxury. They used tweezers . . . and they played games of
" drafts. A number of iron styli showed that they were able to read and
" write, and one decorated tablet of Kimmeridge shale appeared to be of
" the kind used for writing upon, with the stylus, by means of a coating
" of wax spread over it. . . . Some of their houses were, perhaps, covered
" with Roman tegulæ and imbrices, and others were certainly roofed with
" tiles of Purbeck shale. They wore well-formed bronze finger rings, set
" with stones or enamelled, and their fingers were of small size. They
" used bangles of bronze and Kimmeridge shale, and one brooch dis-
" covered, was of the finest mosaic. . . . Also gilt and enamelled brooches,
" some of which were in the form of animals. They used bronze and
" white metal spoons, and the number of highly ornate bronze and white
" metal fibulæ, showed that such tastefully decorated fastenings for their
" dresses must have been in common use. . . . They ate oysters, which,
" considering the distance from the coast, implies a certain degree of
" luxury, though it is possible that the shells may have been used as
" utensils for some purposes. One of the most interesting discoveries
" connected with these people, was the small stature of both males and
" females.[1] . . . . The probability is that both villages were inhabited by
" different classes, and not improbably, they may have been the homes of
" Roman colonists, surrounded by their families, and a bevy of slaves.
" The possibly Roman characteristics, recognized by anthropologists, in
" one round-headed skeleton, may, perhaps, be regarded as favouring this
" view, but the long heads of the majority seem to indicate with great
" probability that the bulk of the inhabitants were of British origin. The
" coins prove that the villages were occupied up to the Constantine period
" (A.D. 306, 337), and Woodcuts certainly, up to the time of Magnentius,
" A.D. 350-353." While in vol. ii, p. 187, Gen. Pitt-Rivers says: " the
" finding of a single interment of the Bronze age in the centre of
" Rotherley points to the probable occupation of the site in pre-Roman
" times."

This masterly account of life in these open British villages during
the Roman occupation is entirely founded upon knowledge gained by the
spade; and it gives us a new measure of understanding for the survey—
though ours may be only superficial—of the other British settlements
that have been enumerated. In several cases, however, notably Gussage
Down, Blandford Race-Down, South Tarrant Hinton Down, and Berwick

---

[1] The measurements made of twenty-eight skeletons averaged 5 feet 2 inches, males, and
4 feet 10 inches, females.

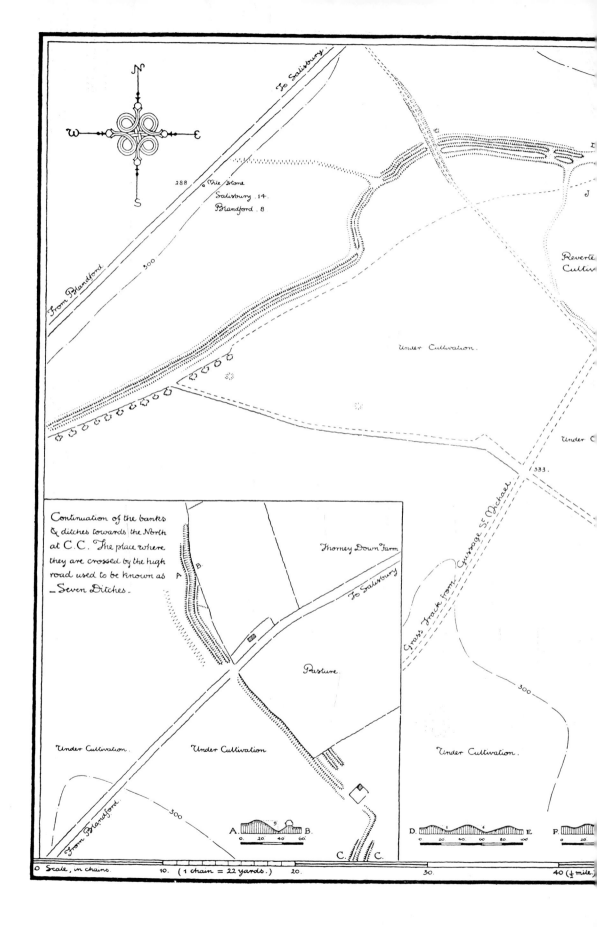

To Salisbury

From Blandford

N
W   E
S

288   Mile Stone
Salisbury . 14.
Blandford . 8

500

Under Cultivation.

Reverte
Cultiv

J

333.

Under C

Under C

Continuation of the banks
& ditches towards the North
at C.C. The place where
they are crossed by the high
road used to be known as
— Seven Ditches.

A
B

Thorney Down Farm

To Salisbury

Cross Track from Cruxcage St. Michael

Pasture

500

Under Cultivation.

Under Cultivation

Under Cultivation.

From Blandford

500

A.   B.
0. 20. 40. 60.

C. C.

D.   E.
0. 20. 40. 60. 80. 100.

P.
0. 10.

O Scale, in chains.   10. (1 chain = 22 yards.)   20.   30.   40 (¼ mile.)

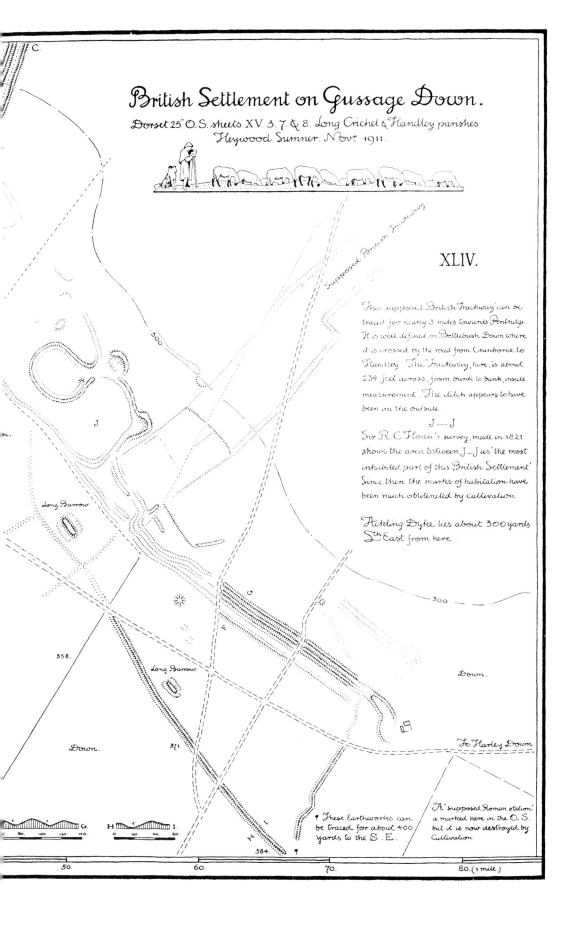

British Settlement on Gussage Down.

Dorset 25" O.S. sheets XV. 3. 7. & 8. Long Crichel & Handley parishes
Heywood Sumner. Nov.r 1911.

XLIV.

This supposed 'British Trackway' can be traced for nearly 3 miles towards Pentridge. It is well defined on Bottlebush Down where it is crossed by the road from Cranborne to Handley. The Trackway, here, is about 254 feet across, from bank to bank, inside measurement. The ditch appears to have been on the outside.

J—J
Sir R. C. Hoare's survey, made in 1821 shows the area between J—J as 'the most inhabited part of this British Settlement.' Since then the marks of habitation have been much obliterated by cultivation

Ackling Dyke lies about 300 yards S.th East from here.

Long Barrow

Long Barrow

358.

37¹

Down.

Down.

To Harley Down

Down.

384.

† These Earthworks can be traced for about 400 yards to the S.E.

'A' supposed Roman station is marked here in the O.S. but it is now destroyed by cultivation

G.   80. 100 120 140.

H   0   20   40.   60.   I.

50.   60.   70.   80. (1 mile.)

Down, their plans are so varied and intricate, that it is not possible to say "ex uno disce omnes." Only systematic excavation could reveal the secret of these tortuous and reduplicated earthworks.

Gussage Down was a site beloved by Sir Richard Colt Hoare. It was here that he placed the Roman station of Vindogladia, and this he claimed as the finest example of a British settlement within the limits of his survey—or without the limits I should say—for Gussage Down is in Dorset, but Sir R. C. Hoare could not here resist crossing the border of his county, and so he includes it in his great work on "Ancient Wiltshire." Now we see less than he saw one hundred years ago. Cultivation has effaced much, and has disconnected the various parts of these complicated earthworks. What still remain only suggest a British settlement; there is no sign of Roman occupation, and until excavation on Gussage Down yields a better solution, Woodyates appears to be the more probable site of Vindogladia.

The present superficial evidence of earthworks marks Gussage Down as having been a great centre in prehistoric times—a pre-eminence which it shares with the settlement on Blandford Race-Down. Warne classes both these settlements as British towns, and he has an elaborate theory of trackway communications radiating from Gussage Down,[1] founded on the banks and ditches which meet there, and which branch out therefrom. It seems doubtful if the theory fits the facts. The superficial measurements of these banks and ditches correspond with those of Grim's Ditch, and the sections that I have excavated in Grim's Ditch, on Gallows Hill, and Knoll Down, disprove the possibility of trackway usage. When used singly these ditches suggest boundary, pastoral purposes, tribal fences; when duplicated and multiplied they suggest defence, either against wolves, or against a vagrant raid. Certainly they now appear to multiply around the settlement, and become single as they trend away from the settlement, and elaborate means of intercommunication between neighbouring settlements seems both unlikely, as a requirement of tribal life, and impossible by means of such ditches. Pastoral boundaries, or, as we should now describe them, fences, appear to have been the main purpose of these entrenchments.

The broad approach that may be traced for two miles across the downs, from Pentridge to Gussage Down, was supposed by Sir R. C. Hoare to be a "cursus," or British race-course. It is about 77 yards wide, and is enclosed on either side with a low bank and an outside ditch, the latter hardly discernible except by the colour and growth of the grass. There

---

[1] "Ancient Dorset," p. 23 *et seq.*

is a somewhat similar broad track between low banks North of Stonehenge. This is 1 mile and 5 furlongs in length, and 110 yards in breadth, but its course is much straighter than the " cursus " between Pentridge and Gussage Down.

The site of Blandford Race-Down is difficult to grasp owing to its size, and to the multiplicity of humps and hollows, banks and ditches, and barrows by which it is covered. The earthworks of many centuries are dimly recorded on this down surface. Long barrows, round barrows, disk barrows, pit-dwellings, boundary ditches, cattle-ways, and cultivation banks may all be found here ; and when the superficial surveyor has done his best to puzzle out this multiplication of earthworks, he will probably question in silence. Some day, perhaps, wise digging for knowledge (on rather a large scale) may wrest the secrets of Blandford Race-Down. Till then, its sequence of occupation must be a matter of conjecture.

Berwick Down is the place-name of a long, high ridge that descends from the Oxdrove Ridgeway, above Ashcombe, to the deep-sunk upland valley of Tollard Royal. A wind-swept site, but once upon a time sought after, as testified by the numerous low banks and shallow ditches that cover the down. It appears to have been partly effaced by cultivation (where the trackway runs from Tollard Royal to the Ferne and Rushmore road), and to have crossed the down from scarp to scarp. It suggests similar occupation to Rotherley (excavated by General Pitt-Rivers, vol. ii), on the adjoining ridge Eastward. I am indebted to Mr. Herbert S. Toms for the plan which delineates a small enclosure—presumably for cattle—connected with this settlement. (See p. 45 *supra*.)

It should be noted that all these British villages are in proximity to barrows, a fact that seems to connect their occupation with pre-Roman times, and, in some cases, even with Neolithic times ; for long barrows adjoin the following sites: Gussage Down, Blandford Race-Down, South Tarrant Hinton Down, Tarrant Hinton Down, Chettle Down, Oakley Down, Woodyates, and Tidpit Common Down ; and there are signs of unproven British settlements near Duck's Nest, Long Barrow, on Rockbourne Down; and on Knoll Down, near Grans Barrow, and Knap Barrow.

The map of Cranborne Chase shows how completely the sites of the old British settlements have been abandoned. They are all now solitary and deserted—only Whitsbury, Woodyates, Ashmore, and Shaftesbury remain as sites occupied both by ancient and modern Britons. As a rule we may generalize, and may say that the British hill-top and upland sites have been abandoned, and that our present-day valley sites mark the choice of the Anglo-Saxon.

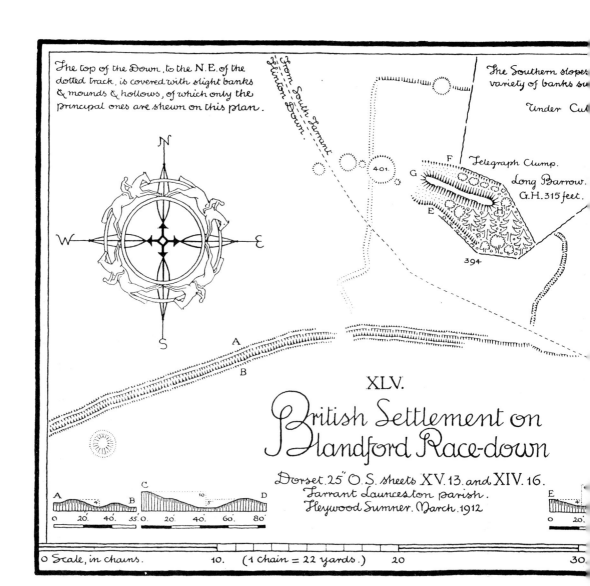

The top of the Down, to the N.E. of the
dotted track, is covered with slight banks
& mounds & hollows, of which only the
principal ones are shewn on this plan.

From South Farrant
Hinton Down.

The Southern slope
variety of banks s

Under Cul

Telegraph Clump.
Long Barrow.
G.H. 315 feet.

401.

F
G
E
H

394

N
W
E
S

A
B

A
B

C
D

XLV.

British Settlement on
Blandford Race-down

Dorset. 25" O.S. sheets XV. 13. and XIV. 16.
Farrant Launceston parish.
Heywood Sumner. March. 1912

A   B   C   10.   D   E

0   20.   40.   55.   0.   20.   40.   60.   80'   0   20'

0 Scale, in chains.   10.   (1 chain = 22 yards.)   20   30.

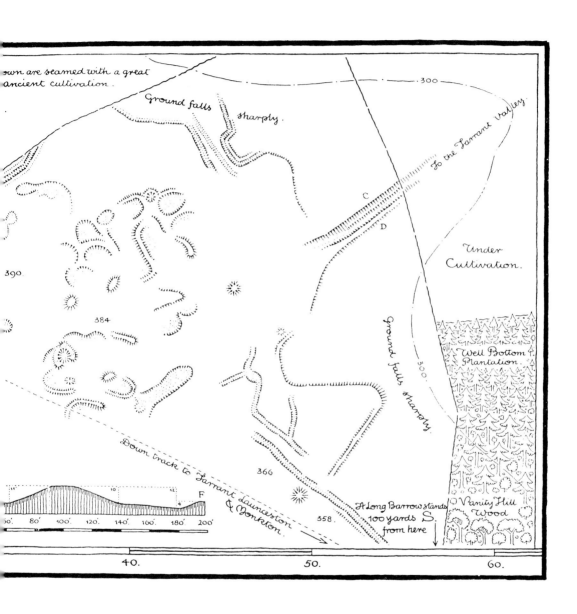

own are seamed with a great
ancient cultivation.

Ground falls
sharply.

─ 300 ─

To the Tarrant Valley

C

D

Under
Cultivation.

390

384

300

Ground falls sharply.

Well Bottom &
Plantation.

F

50' 80' 100' 120' 140' 160' 180' 200'

Down track to Tarrant Launceston & Monkton.

366

358.

A Long Barrow stands
100 yards S.
from here

Vanity Hill
Wood

40.   50.   60.

PIMPERNE LONG BARROW AND BOUNDARY DITCH

Failing water supply undoubtedly changed the habitable conditions of these upland sites. We know, from General Pitt-Rivers' excavation of a Roman well at Woodcuts, that the water-level then stood above its present height. In prehistoric times, during the Neolithic and Bronze age times, it probably stood higher. The valley gravel in the now dry combe bottoms assures us that streams of water once flowed where this gravel is deposited. Accordingly, these ancient British village sites may be assumed to have been better supplied with water—to have been more habitable—than might be supposed from their present conditions. In some cases, on the actual hill-tops, they may have been further helped by mist ponds, but this possible source of supply would have been quite unnecessary for the pastoral enclosures of Tarrant Hinton Down, South Tarrant Hinton Down, Blandford Race-Down, Chettle Down, Chickengrove, Martin Down, Soldier's Rings, Rockbourne Down, and Prescombe Down—all of which sites appear to have been chosen for their well drained position, combined with easy access to water supply.

PIMPERNE LONG BARROW AND BOUNDARY DITCH

THE following extract from " British Barrows, especially those of Wiltshire and the Adjoining Counties," by Dr. Thurnam, " Archaeologia," xliii, may direct the reader to a source of original research on the subject of Long Barrows (and of Barrows generally):[1]

" As a rule long barrows occupy the highest points on the downs, in " situations commanding extensive views over the adjoining valleys, and " so as to be visible at a great distance. Salisbury Plain may be said to be " guarded as it were by a series of such long barrows, which look down " upon its escarpments like so many watch-towers. Others occupy elevated " central spots on the interior of the plain; and some of these—as Ell- " barrow and Knighton-barrow—are well-known landmarks to the hunter " and wayfarer over these extensive and (in winter) dreary downs. Several " of the clusters of round barrows near Stonehenge are grouped around, " or in close proximity to, a single long barrow. On inspecting such a " group as that on Winterbourne Stoke Down, where out of twenty- " seven tumuli we find a single long barrow, or as that on Lake Down, " where to twenty-three circular barrows of various forms we also have

[1] See also " Ancient Wiltshire," by Sir R. C. Hoare, " Celtic Tumuli of Dorset," by Charles Warne, " Stonehenge and its Barrows," by William Long, and " British Barrows," by Canon Greenwell.

" one long barrow, it might at first be thought that the long and circular
" barrows were of the same date, and that the elongated tumulus, as well
" as the variations in the forms of the round barrows, had its origin merely
" in the taste or caprice of those by whom it was erected. Knowing, how-
" ever, as we do, that the examination of the long barrow discloses an
" entirely different method of sepulture, and indicates a much earlier
" epoch than does that of the round barrows, we come rather to regard
" them as the burial places of an earlier race, probably the original pos-
" sessors of the soil, around which the tombs of a later and more cultivated
" people were afterwards erected. As a rule, these tumuli stand apart from
" those of circular form. . . . In by far the greater proportion of long
" barrows, the mound is placed east and west, or nearly so, with the east
" end somewhat higher and broader than the other. Under this more
" prominent and elevated extremity the sepulchral deposit is usually found
" at or near the natural level of the ground; but although this is the general
" rule, a certain proportion depart decidedly from such a system of orien-
" tation, being placed pretty nearly north and south, and this is an arrange-
" ment which I find obtains in about one out of six of our Wiltshire long
" barrows. In this case, as I have found by excavations, sometimes the
" south and sometimes the north end is the higher and broader of the
" two, and covers the sepulchral deposit. They vary in size from one or
" two hundred to three and even nearly four hundred feet in length, from
" thirty to fifty feet in breadth or upwards, and from three to ten or even
" twelve feet in elevation. Along each side of the whole length of the
" tumulus is a somewhat deep and wide trench or ditch, from which trenches
" no doubt a great part, or sometimes even the whole, of the material of the
" mound was dug, but which it is very remarkable are not continued
" round the ends of the barrow.
    " The absence of chambered long barrows in South Wiltshire appears
" to be due to the fact that in those chalk regions there is an absence of
" stone suitable for the construction of chambers."
    The line of the boundary ditch coming from the British settlement
on South Tarrant Hinton Down appears to avoid Pimperne Long Barrow
(see plan). The line of the branch of Grim's Ditch, that runs from Knoll
Down to Blagdon Hill, similarly appears to avoid the wasted long barrow
near Tidpit Common Down (see plan xxxiii) suggesting in both cases
that the long barrows are more ancient than the boundary ditches.

FINIS

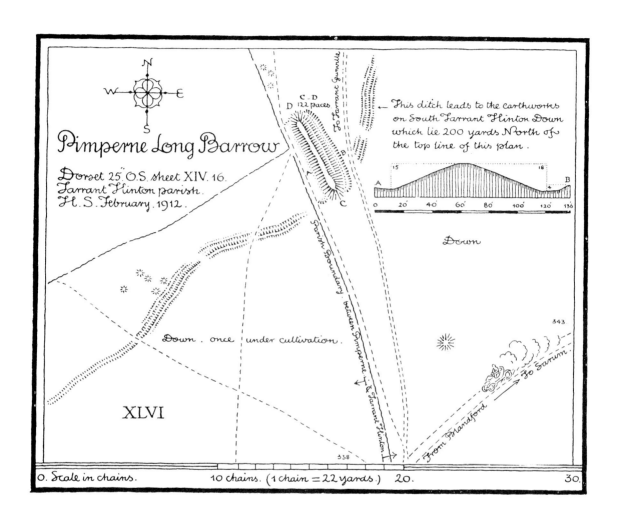

Pimperne Long Barrow

Dorset 25" O.S. sheet XIV. 16.
Tarrant Hinton parish.
H. S. February. 1912.

XLVI

C - D
D 122 paces

To Tarrant Gunville

This ditch leads to the earthworks
on South Tarrant Hinton Down
which lie 200 yards North of
the top line of this plan.

A   B

15   18

20   40   60   80   100   120   136

Down

Parish Boundary between Pimperne

& Tarrant Hinton

Down . once   under cultivation.

338

343

From Blandford

To Sarum

0. Scale in chains.   10 chains. (1 chain = 22 yards.)   20.   30.

APPENDIX

On Pimperne Down (see Map of Ancient Earthworks on Cranborne Chase) close to the cross road leading from Pimperne to Stourpaine, there is a small and wasted fragment of an entrenchment that appears to have formed part of a pastoral enclosure. Potsherds are here thrown up by the moles. The Down shows marks of cultivation banks to the north of this fragment.

At Mountslow (see Map of Ancient Earthworks on Cranborne Chase) there is a small camp of which the entrenchment is slight but fairly well preserved. It is so much concealed by woodland growth that it could only be surveyed when the hazel underwood has been cut. This year, 1913, a small portion was thus uncovered at the western end of the camp, and showed that it ended here with a blunt corner. Until the adjoining underwood has been cut it is impossible to do more than indicate the site, which can scarcely have been chosen for defensive purposes.

Perambulations of Cranborne Chase, 8 Edward I, and 29 Henry III (from "A Chronicle of Cranborne," by Dr. Wake Smart): " 8 Edward I. Pleas of Jurates " and Assize, before Salomon de Ross and his associates Justices itinerant at Wynton, " in the county of Southampton, in the octave of St. Martin, in the 8th and begin- " ning of the 9th year of the reign of King Edward. Gilbert de Clare, Earl of " Gloucester and Hertford, was summoned before the Justices in Eyre, at Sherborn, " in the county of Dorset, to shew by what warrant he appointed to himself free " Chase from the royal way which leads between Shaston and Blandford, on the hill, " on the western side of the said way unto the water of the Stour from the bridge of " Blandford, ascending by the Stour to the water of Sturkel, and by the water of " Sturkel to Shaston; within the precinct of which appropriation, are contained the " villages of Melbury (Melleber), Compton, Fontmel, Iwern Minster, Hanford, " Chylde Hanford, Iwerne Curteney, Ranston (Randulneston), Steepleton, Lazerton, " Ash (Assch), Stourpain, Notford Lock, and Blandford; and thereupon William " de Giselham, who prosecutes for the King, says, that when the said Earl had his " Chase by certain metes and bounds, viz., from Chetell's head to Grimsditch, and " from Grimsditch to Handley, from Handley to Deane, from Deane to Gussich " St. Andrew, to Brandone (Barendown), from Brandone to the head of Stub- " hampton, and by the middle of the vill of Stubhampton to the head of Rithersdene " (Roger Hay's lane), from Rithersdene to the royal way that leads from Blandford " to Shaston, and from that way to Thenerden (Tennersditch), and from Thenerden " to the head of Westwood by the way called Rugwyk (Ridgway), which leads " towards Salisbury to the bounds of Wilts, which lie between Ashmore and Ash- " grove, and so to Stanton, and from Stanton to Mortegresmore, and from Morte-

" gresmore to Singoak, and from Singoak to Sandpit, from Sandpit to the head of
" Longcrofts, and so to Warmer (Larmer), from Warmer to Buckden, from Buckden
" by the metes and bounds which divide Dorset and Wilts, to Chetell's head; the
" Earl appropriated to his Chase beyond those metes and bounds the said villages
" of Melbury, Compton, Fontmel, Iwerne Minster, Chylde Hanford, Iwerne Cor-
" tenay, Ranston, Steepleton, Lazerton, Ash, Stourpain, Notford Lock, and Bland-
" ford, which are out of his Chase, making in them attachments of vert and venison,
" in prejudice of our Lord the King and his dignity, &c. And the Earl came and
" said that there was a perambulation of the metes and divisions of his Chase of
" Craneburn, made in the time when King John was Earl of Gloucester, and that
" afterwards, in the time of King Henry, the father of the king who now is, there
" was made a certain Inquisition by Geoffrey de Langley and Richard de Worthing,
" the Justices appointed for the metes, divisions, and perambulation aforesaid, who
" took the said inquisition at New Sarum, in the 29th year of the same king
" (Henry III.,) on the oath of the underwritten (here follow the names of the Jurors),
" who unanimously said, that these are the metes and divisions by which perambu-
" lation was made of the Chase of Richard, Earl of Gloucester and Hertford; *viz.*,
" from Bulbridge, in Wilton, to Hurdecote (Hurcot), by the water of Nodder to
" the mill of Dyninton to the mill of Tyssebyr (Tisbury), from the mill of Tyssebyr
" to Wycham (Wyke), from Wycham by the water of Nodder to where Semene
" (Sem), falls into Nodder, and so by the water of Semene to Semenehened (Semley),
" from Semenehened to Kingesethe (Kingsettle), near Shaston, from Kingesethe to
" Shaftesbury, namely to Sleybrondesgate, from Sleybrondesgate to the church of
" St. Rombald, from the church of St. Rombald to Gildenhoc (Golden oak), from
" La Gildenhoc to the water of Sturkel, and from the water of Sturkel to the bank
" of Stour, and by the bank of Stour, to the bridge of Hayford, and from the bridge
" of Hayford to the bridge of Blaneford, and from the bridge of Blaneford to the
" bridge of Cranford, and from the bridge of Cranford to Aldwynesbrigg (Julian
" bridge), under Wimborn, and from Aldwynesbrig by the water of Wimborne to
" Waldeford (Watford), from Waldeford to Wichampton, from Wichampton to
" Pontem Petre (Stanbridge), from Pontem Petre to Longam Hayam, which leads
" to Muledich, from Muledich to Kings, from Kings by the way which leads from
" Lesteford through the middle of Estwood,[1] from Lesteford by the Cranborne
" water to La Horewithie (or Horewyethe—Hoarwithy),[2] and from La Horewithie
" to Albelake (Eblake), to Le Horeston, from Le Horeston by the way to the great
" bridge of Ringwood, from the great bridge of Ringwood to the bridge of Ford
" (Fordingbridge), from the bridge of Ford to the bridge of Dinton (Downton),
" from the bridge of Dinton to Aylwardesbrigg (Ayleward's bridge, in Fisherton),
" from Aylwardesbrigg to the aforesaid bridge of Bulbridge, in Wilton."

[1] " There is a difficulty in determining the line of boundary from Stanbridge to Horeston, several
" of the names being lost. Longam Hagam, means probably the long hedge or enclosure which leads
" to Muleditch or Milsditch, these and Kings we cannot indentify; Lesteford, the same probably with
" Letisford mentioned in Domesday survey, cannot be now ascertained; Estwood, the present
" Estworth."
[2] " The word *Hoar*, Celtic and Welsh *Or*, Latin *Ora*, signifies a landmark or boundary of property.
" Thus *Hoarwithy* is a boundary willow tree: *Horeston* a boundary stone " ("Archaeologia," xxv).

INDEX

Ackling Dyke, 48; excavated section of, 49.

Agriculture, Traces of ancient, 6, 28, 29, 33, 40, 42, 43, 44, 50, 69, 74.

Allcroft, Hadrian, his opinion respecting Lydsbury Rings, 14; his description of Knowlton, 47.

"Ancient Dorset," by Charles Warne: camp in Hook's Wood, 32; plans in, 4; plan of oval earthwork within Buzbury Rings, 45; omission of Mistleberry wood camp, 34; his opinion of Bokerly Dyke, 55; of Grim's Ditch, 60; of Knowlton, 46; of trackway communications, 73.

"Ancient Wiltshire," by Sir Richard Colt Hoare: his plans of Castle Ditches (near Tisbury), 18; of Castle Rings, 24; of Chiselbury, 26; of "Whichbury Camp," 20; his record of barrows, 75; his surveys of Clearbury Ring, 27; of Grim's Ditch, 59; of Gussage Down, 56, 73; of Knighton Hill, 43; of Marleycombe Hill, 40.

"Antiquary, The," Papers in, by Herbert S. Toms, on Valley Entrenchments, 43, 45.

"Archaeologia," xliii, "British Barrows," etc., Dr. Thurnam, 75-76, 78.

"Archaeological Journal," 1900, account of Excavations in Hod Hill camp, 14, 19. *See* Austen.

Armitage, Mrs. Ella S., "Early Norman Castles," 50, 51.

Austen, Rev. J. H., "Vestiges of Roman occupation in Dorset," 19.

Badbury Rings, 18, 67.

Badonicus, Mons, its probable site, 19, 21.

Baker, Sir Talbot, excavations on Hod Hill, 14.

Barnes, Rev. William, his opinion of Bokerly Dyke, 55.

Barrow centre of Cranborne Chase, 47.

Barrows near British Settlements, 74; near Oakley Down, 48; Dr. Thurnam on, 75.

Bastion defence at Chickengrove, 40; at Chiselbury, 26; at Hod Hill, 12.

(Bastion—a work projecting outward from the main enclosure of a fortification, enabling its defenders to deliver a flanking fire on their assailants.)

"Belgic Ditches, The," by Dr. Guest, 55, 59.

Berin-byrig, Battle of, 34.

Berm defence at Odstock Copse, 33.

Berm-like projection in the middle ring of Badbury Rings, 19.

(Berm—a projecting path or shelf between the inner and the outer defences of a camp.)

Berwick Down, Earthworks on, 45, 65, 66, 68, 72, 74.

Blackbush Down, "Beaker" found on, 29; British village on, 68.

Blandford Race-Down, British village on, 68, 72, 74.

Bokerly Dyke, 54-57; branch of, so-called in the Ordnance Survey, 57, 59; a typical post-Roman, British, defensive work, 14, 56, 66; excavations in by General Pitt-Rivers, 55; not connected with Great Ditch Banks, 65; opinions as to its origin and purpose, 54, 55.

Borlase, William, "Antiquities of Cornwall," 49.

Bottlebush Down, Cultivation banks on, 50.

Boundary Entrenchments: their convergence towards British settlements, 22, 25, 42, 69, 73; their probable necessity as Tribal landmarks, 61.

Boyd Dawkins, Professor, excavations on Hod Hill, 14.

Breamore Down, 21, 52, 61.

"British Barrows," by Canon Greenwell, 75.

British Museum Guide to the Antiquities of the Early Iron Age, 15.

British village sites, 68-75; conjecture as to the conditions ruling when they were made, 68, 69; results of excavations in three villages by Gen. Pitt-Rivers, 69-72.

Browne, Sir Thomas, xiv.

Browning, Robert, 1.

Burcombe Punch Bowl, Entrenchments near, 63, 65.

Burhs, or Boroughs, their difference from Norman Castles, 50.

INDEX